Charikleia Zouridaki

Security in Mobile Ad Hoc Networks

Charikleia Zouridaki

Security in Mobile Ad Hoc Networks

Trust Establishment for Reliable Data Packet
Delivery in Mobile Ad Hoc Networks

VDM Verlag Dr. Müller

Imprint

Bibliographic information by the German National Library: The German National Library lists this publication at the German National Bibliography; detailed bibliographic information is available on the Internet at http://dnb.d-nb.de.

Cover image: www.purestockx.com

Publisher:
VDM Verlag Dr. Müller Aktiengesellschaft & Co. KG, Dudweiler Landstr. 125 a, 66123 Saarbrücken, Germany,
Phone +49 681 9100-698, Fax +49 681 9100-988,
Email: info@vdm-verlag.de

Produced in USA and UK by:
Lightning Source Inc., La Vergne, Tennessee, USA
Lightning Source UK Ltd., Milton Keynes, UK
BookSurge LLC, 5341 Dorchester Road, Suite 16, North Charleston, SC 29418, USA

ISBN: 978-3-639-01725-0

To my husband, Marek, and our son, Alexandros Georgios.
You are my strength, you are my smile.

Acknowledgments

I would like to acknowledge many people for the realization of this book.

First of all, I wish to sincerely thank my dissertation advisor Dr. Brian L. Mark for all the lessons he has taught me in a number of years. I cannot emphasize enough his devotion to the highest quality of research and his sharp eye for detail, which have made a deep impression on me. He is an outstanding professor and I have been honored to be his student.

I would also like to thank Dr. Kris Gaj for revealing the world of cryptography, and security in general, to me. His classes have made a very strong impact on my scholar and professional choices thereafter. I also thank Dr. Kris Gaj for his careful advice on my research.

My deepest gratitude and love goes to my parents, Georgios and Nineta Zouridaki for their endless love, understanding, support and interest in my work and life. They have defined my existence in wonderful ways and their confidence in me and my choices has been the point of reference in my life. Similarly, I am grateful to my brothers, Konstantinos and Pavlos, for being such fulfilling and significant parts of my life, as well as my grandparents and my closest family. A special soul, Orfeas, will warm my life forever. Additionally, I would like to thank my parents-in-law, Mieczyslaw and Marta, for welcoming me into their family and their sincere interest in my wellbeing. It goes without saying, that I also owe my warmest thanks to all my true friends, for being so caring, humorous and bright, no matter what.

But most of all, I wish to express my endless love and gratitude to my husband, Marek Hejmo, and our son Alexandros Georgios. Marek, our wedding, which we celebrated while we were both pursuing our PhD degrees, our marriage and the birth of our son, are the highlights of my life. Alexandros Georgios, your existence has added another wonderful dimension to my life. I cannot wait to spend each and every day of the rest of my life with you. You are always inspiring me with your love.

Table of Contents

Chapter 1: Introduction

1.1 Motivation and Problem Statement

In mobile ad hoc networks (MANETs), a source node must rely on other nodes to forward its packets on multi-hop routes to the destination. Secure and reliable handling of packets by the intermediate nodes is difficult to ensure in an ad hoc environment. The lack of infrastructure, the limitations of the wireless channel, and the limited resources of the mobile nodes [1,2] make it challenging to guarantee reliable packet delivery in the presence of malicious nodes acting as intermediate hops. Additionally, most of the currently proposed routing protocols for mobile ad hoc networks assume that the network nodes are protocol compliant and forward packets as expected. Malicious nodes can exploit the network protocols, which do not incorporate mechanisms to check the compliance of the nodes. Malicious nodes may choose to drop packets in order to preserve their limited resources and battery life.

In a network, a source node typically can choose a route from among multiple paths to forward packets to a destination node. To maximize the chance that the packets are forwarded correctly, the most "trusted" path should be selected. Hence, there is a need to evaluate the trustworthiness of nodes with respect to their reliability in packet delivery. A trust establishment scheme that relies on network observation data is needed to evaluate the *trustworthiness* of the network nodes in the ad hoc environment. We adopt the definition of *trust* given by Morton Deutsch [3], as it is the most widely accepted definition of trust. This definition is acceptable to the extent that the basic structure of a trusting choice is shown:

"Trusting behavior occurs when an individual perceives an ambiguous path, the result of which could be good or bad, and the occurrence of the good or bad result is contingent on the actions of another person; finally the bad result is more harming than the good result is beneficial. If the individual chooses to go down that path, he can be said to have made a trusting choice, if not, he is distrustful."

We refer to *trust* and *trust establishment* in terms of reliable data packet delivery in mobile ad hoc networks. Our goal is to develop a trust establishment framework to evaluate the trustworthiness of the network nodes for reliable data packet delivery in the ad hoc environment.

1.2 Trust Establishment for Reliable Packet Delivery

To improve the reliability of packet delivery, we propose a trust establishment scheme, which we call Hermes [1], that enables a source node to route packets over more "trustworthy" intermediate nodes. In the proposed scheme, each node assigns a "trustworthiness" metric to each of its neighbor nodes based on direct observations of packet forwarding behavior. The concept of trustworthiness is extended to the notion of an "opinion" that a node has of any other node. The opinion metric can be applied in a various network settings to improve packet

---

[1]In Greek mythology, Hermes was the trusted messenger of the gods.

1

delivery performance. In particular, the opinion metric can be incorporated into MANET routing protocols such as DSR [4] or AODV [5] to route packets on more "trusted" paths or can be used as a criterion to maintain network membership [6].

Our proposed trust establishment scheme makes use of a Bayesian approach similar to that used in [7]. In the Bayesian approach, trust values are computed under the assumption that they follow a beta probability distribution. The parameters of the beta distribution are estimated by accumulating empirical observations of packet forwarding behavior. A *trust* metric can then be derived from the parameters of the beta distribution. Our approach to trust evaluation differs from that in [7] in that we derive an additional parameter called *confidence*, which characterizes the statistical reliability of the computed trust metric.

The notion of maintaining two metrics, trust and confidence, is also considered in [8]. In [8], the trust and confidence metrics assigned to nodes are extended to paths via a semi-group approach. In contrast, we propose a new metric, called "trustworthiness," which combines the trust and confidence metrics. The trustworthiness metric is used to formulate the more general "opinion" metric, which can be incorporated into routing protocols in a transparent manner. We present a windowing scheme to systematically expire old observation data in order to maintain the accuracy of the opinion metric.

1.3   Main Challenges to Establish Trust in Mobile Ad Hoc Networks

Mobile ad hoc networks have unique characteristics which make the task of establishing trust for reliable data packet delivery especially challenging. Below we have summarized the difficulties and issues for establishing trust in MANETs.

- **Lack of Fixed Infrastructure.** Unlike the Internet, in MANETs, there is no clear distinction between routers and traffic sources. Mobile nodes are called upon to act both as routers and traffic sources, even though typically they have limited resources in terms of processing power, memory and most importantly, battery life. A mobile node depends on other mobile nodes to forward its packets. Selfish nodes may choose to avoid forwarding packets, or equivalently to drop packets, in order to preserve their limited resources.

- **Unpredictable Environment Characteristics.** Wireless media is very unpredictable. The transmitted signals face difficulties such as signal fading, interference, multi-path fading and packet collisions. Consequently, a node might be required to forward the same packet multiple times, thus consuming its resources.

- **Limited Bandwidth.** The wireless channel provides limited bandwidth, which places a high premium on signaling overhead.

- **Node Mobility/Intermittent Connectivity.** Node mobility results in a dynamic network topology. Mobile nodes interact and depend on other nodes to forward their packets while they move in the network. Due to node mobility, mobile nodes need to establish trust in terms of reliable packet delivery for every new node with which they interact. Additionally, since nodes move, they interact with each other when they happen to be in each other's radio transmission and reception range. This interaction may be infrequent, making the trust establishment task difficult, as incomplete information might be collected.

- **Limited Energy/Battery Life.** Mobile nodes have restricted battery life and transmission power. A trust establishment scheme that does not introduce excessive communication overhead in terms of number and size of messages is desirable.

- **Limited CPU/Memory.** The nodes of an ad hoc environment typically have limited processing power and memory; yet, they are called upon to act both as routers and traffic sources. It is essential to develop a trust establishment scheme, which minimizes the storage and processing power requirements for the mobile devices.

1.4 Research Contributions

The objective of this book is to develop a complete and quantitative framework for trust establishment with respect to reliable data packet delivery for mobile ad hoc networks. The main contributions of Hermes can be summarized as follows:

1. a scheme for evaluating trust and confidence with respect to packet delivery based on empirical observations;

2. a scheme for mapping trust and confidence into a "trustworthiness" metric and its extension to an "opinion" metric;

3. a windowing scheme to improve the fidelity of the opinion metric;

4. a scheme for accumulation of first-hand trust information (obtained independently of other nodes) for non-neighbor nodes;

5. a scheme that accelerates the convergence of trust establishment procedures, yet is robust against the propagation of false trust information by malicious nodes;

6. mechanisms to make the trust establishment framework robust to Byzantine behavior, i.e., arbitrary, deviant behavior;

7. a punishment policy that discourages selfish node behavior;

8. an authentication scheme for both data packets and control packets used for trust establishment;

9. a security evaluation of the scheme based on a probabilistic attacker model to characterize the security properties of the trust establishment procedures

10. a performance evaluation of the scheme providing the overhead incurred by the scheme, its accuracy and convergence;

11. an approach to incorporate the opinion metric into ad hoc routing protocols to improve reliable packet delivery;

12. computer simulation results to demonstrate the effectiveness and the performance properties of the proposed scheme in a variety of scenarios involving nodes that are malicious both with respect to packet forwarding and trust propagation.

In principle, the Hermes framework can be applied to any MANET on top of the routing protocol. The topic of secure routing for MANETs has been studied extensively in recent years [9–14]. In particular, Hermes could be applied to one of the secure routing protocols proposed in the literature [13,15]. A full treatment of trust-aware routing is beyond the scope of the present book.

1.5 Book Outline

The remainder of the book is organized as follows. Chapter 2 presents the literature review of the state-of-the-art. Chapter 3 proposes a trust establishment framework, which enables a given node to determine the "trustworthiness" of other nodes with respect to reliable packet delivery. Chapter 4 extends this trust establishment framework to address an attacker model where nodes can exhibit malicious behaviors independently, i.e., failure to forward packets is independent of the honesty with which trustworthiness values are propagated about other nodes. Another extension is that of a mechanism for deriving trustworthiness values for non-neighbor nodes based on first-hand information from acknowledgements, as opposed to relying on second-hand recommendations alone. Chapter 5 discusses new mechanisms to make our trust establishment framework robust to Byzantine behavior, i.e., arbitrary, deviant behavior, which disrupts packet transmission in the network and introduces a punishment policy that discourages selfish node behavior. Finally, the book is concluded in Chapter 6.

Chapter 2: State-of-the-Art Literature Review

Trust has been a subject of research in the fields of sociology, philosophy, socio-psychology, economics, computer science and engineering for many years. Many researchers have attempted to define and explain trust, as well as develop schemes that make use of the notion of trust. In this Chapter, we present a comprehensive survey of the relevant literature to be found on trust in computer networks and MANETs in particular.

2.1   Trust Establishment in Wired Networks

Marsh in [16] formalizes the notion of trust and implements his formalism in an intelligent artificial agent. Although the sociological foundations of his model are strong, Marsh tries to incorporate all aspects of social trust and introduces a large number of variables into his model; these variables represent abstract notions such as *risk* and *competence*. This makes his model large and complex. Additionally, the application of one variable onto another introduces ambiguity into the model.

In [17], Abdul-Rahman and Hailes propose a trust model for virtual communities that is based on real-world social trust characteristics, utilizing a reputation mechanism. Agents are able to exchange reputational information through recommendations. An agent places *direct trust* in another agent's trustworthiness within a *certain context* to a certain *trust degree*. The trust degrees assigned are *very trustworthy, trustworthy, untrustworthy, very untrustworthy*. The choice of four trust degrees is questionable. Actually, the authors acknowledge the ad hoc nature of the trust (and experience) degrees, expressing the need for a more concrete representation for these metrics. The model is quite complicated and involves many concepts. This model is referenced by [18] as a model most suitable for less formal, provisional trust relationships. One limitation of this trust model when specifically applied to mobile ad hoc networks is the requirement that recommendations about other entities are passed. Subsequently, the handling of false or malicious recommendations should be supported via some out-of band mechanism or a mechanism that evaluates the truthfulness of the recommendation. Another limitation is the requirement that new agents be equipped with a number of trusted entries before they enter the network. No framework is presented to accommodate this requirement.

In [19], Yu and Singh present a complex model, which makes use of notions such as reputation, gossip, propagation of reputation rating, referral chains, queries to assign the trust level, etc. In [20], the authors present another reputation system that collects, aggregates, and distributes feedback about an entity's former transactions. In contrast, [21] provides a classification of trust, a categorization of trust types, and discusses trust direction and symmetry. We remark that in this book trust is defined in terms of a specific context, that of reliable data packet delivery.

Reference [22] presents a trust model for peer to peer networks overlaying social networks. The main idea is that using a prior knowledge or relationship perhaps from a real life social context, may be key to mitigating the effects of misrouting. The objective of this paper is to route packets among computers, not mobile nodes. As a result, the authors choose the route of maximum routing opinion for forwarding. This is not the optimal solution for mobile ad hoc networking. Querying constantly the same nodes for packet forwarding would drain their resources. We propose that a route should be chosen with probability proportional to the routing opinion. We also identify the need to discourage node selfishness so that all network nodes participate in the forwarding functions.

2.2   Trust Establishment in MANETs

As mentioned in the Introduction, our proposed trust evaluation framework is based on a Bayesian approach similar to the one presented in [7]. Key differences, however, are that our framework incorporates (1) the notion of statistical confidence associated with a trust value and (2) a windowing mechanism to improve the accuracy of the trust metric by allowing past observation data to expire. In contrast, in [7], the new observation data is weighted less as the amount of observation data collected increases over time, since observation data does not expire. In [7], a threshold on the trust value is used to classify a node as either (*normal* or *misbehaving*), whereas in this book the trust value is used to make relative comparisons among nodes. While [7] discusses trust establishment from a high level prospective, our goal is to develop specific algorithms to establish trust in order to improve the reliability of packet delivery. The CONFIDANT protocol is presented in [23], which is based on the framework proposed in [7].

The notion of confidence was proposed in [8,24] and a semi-ring approach was suggested to evaluate trust and confidence along network paths. In our approach, however, we map trust and confidence into a new metric, called "trustworthiness," which can more transparently be incorporated into network decisions such as route selection. Furthermore, our framework deals directly with the issue of collecting evidence from the network. Finally, the issues of the assignment of the initial trust values, trust updates and the definition of *bad* nodes are not specified in [8,24]. Similarly, in [25,26] no method is presented for the collection of empirical evidence, whereas trust is entropy-based to measure uncertainty. Additionally, the sensitivity to node behavior changes after many observations as defined by a forgetting factor, depending on how fast the behavior of "agents" changes.

In [18], a trust model is presented that allows the evaluation of the reliability of the routes, using only first-hand information. On the other hand, our approach to trust evaluation incorporates third-party information to derive the notion of an opinion that a given node has for any other node.

The authors of [27] present a high-level framework for generation, revocation and distribution of trust evidence and demonstrate the significance of estimation metrics in trust establishment. The authors argue that a large body of trust evidence has to be generated, stored and protected across the network nodes, routed where needed and evaluated speedily to validate dynamically formed trust relations. Swarm Intelligence is introduced as a tool for evidence distribution. A mechanism for trust evidence dissemination based on a model of ant behavior is proposed in [28] along the lines suggested in [27]. In contrast, our work focuses on developing metrics and mechanisms for establishing trust with respect to the objective of reliable packet delivery. In [27,28], a trust model of two components is presented consisting of a trust evidence distribution system (discussed in [28]) and a trust computation scheme (proposed in [29]). In the trust evidence distribution system, evidence is presented in the form of trust certificates. A limited number of *signers* is present in the network, whose public keys are well known and authenticated. Due to the limited number of signers, the public key authentication is proposed to be done off-line, before the setup of the network. When a certificate in needed, "ants are sent out". Once a forward ant, which represents a request message, "finds" the required certificate, a backward ant, which represents a reply message, is generated. A nonlinear reinforcement-learning rule proposed in [30] is used. All nodes have to agree on a common network view.

The trust computation schemes of [29, 31] assume the existence of pre-validated trusted nodes, and off-line key authentication, as in [28]. In [29], during the voting process, pre-trusted nodes or trusted nodes who received their certificates from pre-trusted nodes, vote for the trustworthiness of the neighboring nodes. All nodes have to agree on a common network view, unlike in our approach, where each node is allowed to form its own view of the network. Moreover, we do not assume the existence of trusted nodes in the network. Our approach does not rely on the exchange of certificates to establish trust. The main result of [31] is the derivation of a lower bound on the number of pre-trusted nodes that are required in order to form a "fully trusted" network. We remark that the certificate revocation problem is left as an open issue in [29, 31].

In [32], a set of trust values is assigned to nodes in the network. The AODV routing protocol is modified such that a node applies different encryption keys to arriving packets depending on the trust value of the node and the security level required by the packet. However, the issue of how to compute the trust values assigned to the nodes is not addressed.

In [33], a framework for stimulating cooperation in MANETs is proposed. The approach is based on a credit system for packet forwarding. A tamper-resistant hardware module, called a security module, is assumed in each node. The protocol implemented requires each packet generated or received for forwarding to be passed to the node's security module. The security module maintains a counter, called a nuglet counter, which is decremented when the node wants to send a packet as an originator, and incremented when the node forwards a packet. The main focus of [33] is to calculate the maximum number of packets that a node can send as an originator and the maximum number of packets that a node has to forward, as well as to investigate the node behavior in order to allow a node to achieve its theoretical maximum values. In contrast to our work, [33] assumes that nodes cannot be trusted to forward packets, if tamper-resistant hardware is not in place. However, the feasibility of a tamper-resistant hardware is debatable [34, 35]. Another unresolved issue is that nodes at the edge of the network may starve because only a small number of nodes might need them for packet forwarding.

The goal of collaboration is also pursued in [36], which proposes a trust management model, whereby each node carries a portfolio of credentials, which it uses to prove its trust-worthiness. Each credential lists the agent that trusts another agent at a predefined level to carry on a service of a known context at a given time. The degree of knowledge in the trust experience is also expressed. However, the choice of the trust formation, dissemination, aggregation and extraction functions is not developed fully for a specific application. Furthermore, the user of a device is said to be able to select the "trust profile" that better describes his natural disposition, from a list of available profiles that are offered to him by the trust management framework, when no details are provided for the trust profile and its parameters, as noted by the authors.

In [37], a trust framework is proposed for the purpose of establishing a set of group keys.

Nodes are organized into trust-based clusters called physical-logical trust domains and are allowed to enter or leave these domains. Trust updates and the frequency with which network information is collected are not addressed. In [38] a collaborative reputation mechanism (CORE) is presented, where a global reputation table is built to combine the different values for reputation calculated for different functions. In this approach, more relevance is given to past observations, which is vulnerable to an attack where a node builds up a good reputation before misbehaving. In [39], Rebahi et al. propose a reputation-based trust model, where every node monitors the behavior of its neighborhood and upon detecting an "abnormal action", it broadcasts this information. Other key differences with our framework are that the proposed solution in [39] collects observations only for neighbors nodes and defines reputation (equivalent to trust) as a function of the number of packets forwarded by a node over the overall number of packets sent for forwarding. On the contrary, we adopt a Bayesian approach to calculate trust, which captures the confidence on the trust value.

The goal of [40] is to provide a secure and distributed authentication service with the use of clustering. As in [28, 29, 31] pre-trusted nodes are assumed, which can issue certificates to nodes that they consider trustworthy. These works do not address the issue of how nodes should monitor the behavior of their group mates and obtain their public keys and how malicious nodes are identified. The goal and approach of the dynamic trust model presented in [41] are very different. In [41], a high-level description of an intrusion detection system (IDS) implemented on each node is described with the aim of detecting Sybil and blackmailing attacks. Reference [42] introduces a cluster-based trust evaluation scheme, in which each cluster head issues a trust value certificate for the cluster member nodes. In this scheme, a cluster is formed based on the trust values of the neighbors nodes, and a cluster head has the highest trust value.

The main focus of [43] is to introduce a framework of three components that prevents selfish behavior, and performs node identification and secure routing. In [43], the authors present a method for node identification through certificate generation, revocation and evaluation. Secure routing is investigated in [44], whereas [45] discusses how "sensors" can detect selfish behavior. In this approach, a number of "local ratings" (equivalent to recommendations) is averaged to calculate the global rating, which determines network membership. More precisely, a set of thresholds dictates when a membership is revoked. Activity-based overhearing is used to differentiate node misbehavior from non-malicious node fault, that is when a node sends a packet to another node and cannot detect a forwarding of the packet by the relaying node, this is esteemed a selfish behavior only when there has been a recent activity by this node. However, this is an arguable assumption, as a non-malicious malfunctioning node may be generating traffic in an arbitrary manner. We do not differentiate between malicious and non-malicious forwarding misbehavior, as our goal is not to revoke the node membership, but to avoid using a node as relay as long as it does not forward traffic. A windowing mechanism that expires old empirical evidence and probabilistic route selection is proposed for improving the accuracy of the trust metric.

In [46], the authors propose a trust diffusion protocol. The main focus of the paper is the presentation of the definitions of "trust mask" (equivalent to trust), "trust index" (equivalent to confidence), "external trust knowledge" (equivalent to recommendations), "trust index of external trust knowledge" (equivalent to confidence on recommendations) and a method for trust evolution. However, the procedures for trust updates are not presented, e.g., recommendations are "simply transmitted through the network by nodes at each interaction". The paper emphasizes the importance of recommendations, while only negative recommendations are used for trust updates. Another interesting choice is that at system initialization, the confidence is considered 0.5 on a scale between 0 and 1.

The approach presented in [47] differs in that it attempts to identify "critical" nodes in the network. A critical node is a node whose failure or malicious behavior disconnects or significantly degrades the network performance. Once identified, a critical node can be the focus of more resource intensive monitoring. As the authors note, any kind of network diagnosis depends on the mobility of the nodes, and therefore they aim in determining an approximation of the network topology, based on which the critical nodes are identified. Reference [48] presents ADOPT (ad hoc distributed OCSP for trust), which provides certificate status information to the nodes of a MANET, and suggests the use of a trust assessment framework named ATF (ad hoc trust framework). ADOPT is deployed as a trust-aware application that provides feedback to ATF, which calculates the trustworthiness of the peer nodes' functions and helps ADOPT to improve its performance by locating valid certificate status information.

2.3 Trust Establishment in Terms of Authentication for MANETs

Trust can also be granted through an authentication framework, whereby certificates can be issued by either of the following:

- a distributed public key infrastructure [9, 49]

- users based on the establishment of chains of trust, in a decentralized way [50–52]

- an out-of-band mechanism [53, 54] or over a location-limited channel [55, 56].

Trust establishment through an authentication framework involves the use of schemes for certificate issuing, revocation and revalidation, which are challenging tasks in the ad hoc environment. However, even nodes certified to be trusted for packet forwarding may exhibit malicious behavior. An authentication mechanism addresses the *outsider* threat, but cannot solve the *insider* threat. The outsider threat comes from entities that do not participate in the authentication process and therefore are excluded from the system, whereas the insider threat comes from authenticated entities that are authorized to be part of the system. Insiders are entities considered benign and protocol compliant, but they may choose to exploit protocols that do not incorporate mechanisms for node-compliance checking.

The currently proposed routing protocols for mobile ad hoc networks assume that the authorized network nodes are protocol compliant. The implementation of a trust establishment scheme that relies on network observation data, rather than an authentication mechanism, is needed to evaluate the trustworthiness of the network nodes. Trust establishment schemes that rely on network observation data were discussed in Section 2.2.

In [49], we present a practical distributed certification authority (CA) based public key infrastructure (PKI) scheme for mobile ad hoc networks based on elliptic curve cryptography that overcomes the challenges of the ad hoc environment. In this scheme, a relatively small number of mobile CA servers provide distributed service for the mobile nodes. The key elements of our approach include the use of threshold cryptography, cluster-based key management with mobile CA servers, and elliptic curve cryptography. The proposed scheme is resistant to a wide range of security attacks and can scale easily to networks of large size.

The Hermes framework presented in this book and the distributed PKI scheme proposed in [49] are complementary and form the basis for a complete framework for secure key distribution and trust establishment in terms of reliable data packet delivery for mobile ad hoc networks.

2.4 Intrusion and Anomaly Detection Systems for MANETs

Recently, several intrusion detection systems (IDS) and anomaly detection systems (ADS) have been proposed for mobile ad hoc networks [57, 58]. Li and Wei [59] present a survey of many IDS and give valuable guidelines on selecting intrusion detection methods for the ad hoc environment.

An IDS uses a defined set of rules or filters that have been crafted to catch specific, malicious events. This is often referred to as misuse or signature detection. When an ADS is used, the profile of usual or normal behavior in place is compared against real-time events. Anything that deviates from the baseline is logged as anomalous. There are two types of ADS: statistics-based and specification-based. The statistics-based ADS employs statistics to construct a point of reference for system behavior. The specification-based ADS depends less on quantitative metrics and more on human observation and expertise. This method uses a logic-based description of expected behavior to construct a profile. It requires a language or standard that can be interpreted by the ADS. Using such a syntax, the administrator constructs a list similar to the rules and signatures used by an IDS. But instead of looking for a predefined intrusion, these rules block any event falling outside the baseline. It is a concept similar to a "Deny-All" firewall, where rules are constructed to block everything except traffic that corresponds to a predefined profile. An IDS can only catch events that it has been set to look for. Anything outside of this list will not be recognized. An ADS, in turn, has the potential to detect new, unknown, or unlisted events.

Current IDS and ADS tend to be heavy in terms of protocol overhead and are not easily adapted for the ad hoc environment. Moreover, the current IDS or ADS cannot determine whether nodes forward correctly the packets received for forwarding. Receiving or forwarding packets is not considered malicious behavior for IDS and ADS that are specification-based. ADSs that are statistics-based check the statistics of the network traffic including the number of packets sent and received. Nonetheless, the number of packets sent by a node include the packets originated and forwarded by the node. Therefore, statistics-based ADS cannot determine whether nodes reliably forward the packets received for forwarding.

Additionally, the output of an intrusion or anomaly detection system is the detection of the intrusion. Detecting the intrusion, in our case, means identifying nodes that do not forward the packets received for forwarding. Detecting the intrusion is only a part of the Hermes framework. Our proposed method for detecting which nodes tend to drop the packets received for forwarding is discussed in Sections 3.4 and 4.3.

2.5 Byzantine Detection for MANETs

Byzantine Detection protocols [60, 61] aim to discover the identity of the "failed" links, i.e., the links that drop packets either because of a network fault or malicious behavior. Byzantine Detection protocols do not necessarily exhibit Byzantine Robustness [1]. In other words, Byzantine Detection protocols do not necessarily operate correctly in the face of Byzantine failure. Most often, the action taken is to remove the link that exhibits Byzantine failure, or the failed link, from the topological map of the source. The impact of mobility on Byzantine Detection protocols is profound, since links are evaluated while mobility changes links over time.

Byzantine Detection protocols, in order to pinpoint links to blame for delivery failures, use acknowledgements (ACKs) generated by the destination and intermediate nodes, timeouts, fault announcements (FA), and cryptographic tools for message authentication. Byzantine Detection protocols can become quite complex, depending on the precision with which they perform failure detection (cf. [61]).

The main differences between a Byzantine Detection protocol and Hermes, or more generally any trust establishment framework, are as follows: (1) Byzantine Detection protocols are routing protocols, which also perform the task of failure detection. Hermes can be applied to any routing protocol; (2) Byzantine Detection protocols aim to detect link failures, whereas Hermes aims to assign trust values to the network nodes. The assigned trust values can be used in trust-aware routing, for key establishment (cf.[37]), etc.; (3) most Byzantine Detection protocols erase faulty links from the topological map of the source. Hermes does not isolate faulty nodes, but uses a windowing mechanism in order to maintain the accuracy of the trust placed on a node, allowing the detection of changes in node behavior. Thus, nodes that misbehave due to a software malfunction or failure, can participate in the network after they recover.

2.6 Summary

The main point of this Chapter are summarized as follows:

- The trust establishment schemes designed for wired networks are complex, often heavyweight, and most importantly, they do not take into consideration the unique characteristics of the wireless environment and the limitations of the mobile devices.

- Most of the proposed trust establishment frameworks for MANETs tend to be at a high level of abstraction and do not address important issues such as accumulation of trust evidence from observation data and computation of a trust metric. Some proposals assume the existence of pre-trusted entities, or tamper-resistant hardware, while others take different approaches in major design issues, for example, the use of recommendations.

- A trust establishment scheme that relies on network observation data, and an authentication mechanism are complementary entities. The focus of this book is on trust establishment from network observation data and propagation of trust.

---

[1]Byzantine Robustness schemes exhibit correct behavior in the presence of arbitrarily malfunctioning nodes, but do not perform Byzantine detection.

- IDS or ADS cannot detect whether a node correctly forwards a packet that has been sent to it as an intermediate hop.

- Trust establishment differs from Byzantine detection in that the classification of nodes is fine-grained. Moreover, trust is a dynamic quantity that may change over time. Byzantine detection aims to identify link failures, and there is a risk that links may be misclassified as "bad" and hence removed from the network. Trust establishment does not classify nodes as good or bad per se, but assigns trust values, which can be used to influence network decision algorithms, such as routing.

Our goal is to develop a quantitative framework for trust establishment for reliable data packet delivery over multi-hop routes in the presence of potentially malicious nodes. Our proposed Hermes framework attempts to address the open research issue of trust establishment for reliable data packet delivery thoroughly. Hermes considers the issues of accumulation of empirical evidence, expiration of past information, trust evaluation from first-hand observations, information sharing among nodes, formulation of opinion for every network node, and discouragement of selfish node behavior.

Chapter 3: A Quantitative Trust Establishment Framework

In this chapter, we propose a quantitative trust establishment framework for MANETs, which aims to improve the reliability of packet forwarding over multi-hop routes in the presence of potentially malicious nodes. Most of the material in this chapter also appears in [62,63].

An overview of the Hermes trust establishment framework is provided in Section 3.1. Section 3.2 describes a methodology for evaluating trust between two neighbor nodes from first-hand observation data. We use the term *trustworthiness* to denote this notion of trust. Section 3.3 extends the trust evaluation scheme to a general pair of nodes. We use the term *opinion* to denote this extension of the trustworthiness concept. The issues involved accumulating trust information from first-hand observation data are treated in Section 3.4. In Section 3.5, the security properties of Hermes are discussed in terms of a probabilistic attacker model. The application of the opinion metric to realize "trust-aware" ad hoc routing is discussed in Section 3.6. Section 3.7 evaluates the performance of our scheme. The accuracy and convergence of Hermes, as well as results from simulation experiments are presented, which demonstrate the key properties of the proposed trust establishment scheme. Finally, this chapter is concluded in Section 3.8.

3.1 Overview

In this Section, we provide a high-level, conceptual overview of the Hermes trust establishment framework as illustrated in Fig. 3.1.

3.1.1 Basic Definitions and Terminology

Two nodes are said to be *neighbors* if they are in the radio transmission and reception range of each other. Due to the broadcast nature of the wireless medium, a given node can collect first-hand information about the packet forwarding behavior of its neighbor nodes by snooping all received frames at the MAC layer and recording packet delivery statistics. The accumulation of observation data is discussed in Section 3.4. The packet statistics can be used to compute a pair of values associated with each neighbor node, which we call *trust* and *confidence*.

We define *trust* as a value in the range $[0,1]$, which indicates the degree to which a given node believes that its neighbor node behaves normally or can be "trusted" to forward the packets it receives for forwarding. The trust value is a statistic computed from first-hand observation data. A value of trust close to 0 implies that the neighbor is likely to be unreliable in forwarding packets, whereas a trust value close to 1 implies that the neighbor is likely to be reliable in its packet forwarding behavior. We define the *confidence*, c, associated with a given trust value t, as a value in the range $[0,1]$, which indicates a measure of statistical dispersion in the trust value. The pair of values (t,c) characterizes the degree of belief that the neighbor is reliable with respect to packet delivery and the statistical confidence in this

13

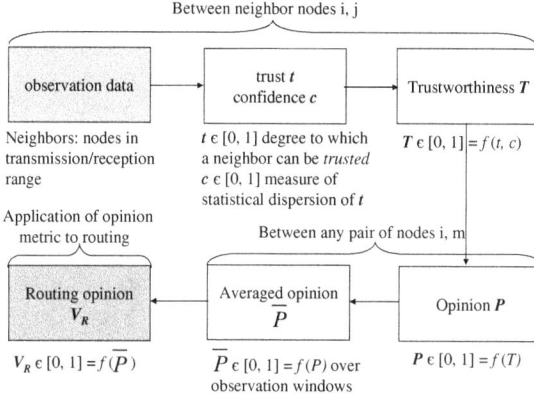

Figure 3.1: Overview of Hermes.

belief. As discussed in Section 3.2, we apply a Bayesian framework to calculate trust and confidence values.

While it is useful to characterize empirical trust via the pair of values (t, c), it is often more convenient to characterize trust in terms of a single value in order to make decisions or take action with respect to the neighbor. Therefore, we define the concept of a *trustworthiness* value T, which is a function of the pair (t, c). The value of T also lies in the range $[0, 1]$. The issue of how to map (t, c) into T is discussed in Section 3.2.3.

Thus far, we have defined the notions of trust, confidence, and trustworthiness only with respect to a node and its neighbor. We now define a more general notion called *opinion*. A given node i can formulate an opinion for any other node m in the network. If nodes i and m are neighbors, then the opinion that node i has for m equals the trustworthiness of node i for m. Otherwise, if i and m are not neighbors, the opinion that i has for m is determined as a function of the trustworthiness values along network paths from i to m.

3.1.2 Notation

Suppose that node i collects observation data related to the packet forwarding behavior of its neighbor j over a time window W. At the end of the window W, node i computes a pair of trust and confidence values with respect to j, which are denoted by $t_{i,j}$ and $c_{i,j}$, respectively. The pair of trust and confidence values can be mapped to a trustworthiness value denoted by $T_{i,j}$. The *opinion* that i has for another node m is denoted by $P_{i,m}$. In case $m = j$, we have that $P_{i,m} = T_{i,j}$. Otherwise, the computation of $P_{i,m}$ requires the propagation of trustworthiness values along network paths from i to m as discussed in Section 3.3.

More generally, the trust values are updated at the end of observation windows belonging to a sequence. Let W_k denote the kth time window. We denote by $t_{i,j}^k$ and $c_{i,j}^k$, the trust and confidence values that node i has for node j, respectively, computed at the end of window W_k. Similarly, $P_{i,m}^k$ denotes the opinion that node i has for node m at the end of W_k. An *averaged* opinion $\bar{P}_{i,m}^k$ can be calculated based on the opinion values computed in previous time windows (see Section 3.4.4). To apply the notion of option to improve routing, the notion of *routing opinion*, V_R, is defined to represent the opinion that a source node s has for a route R, and is used to select the most trustworthy route among a set of alternative routes from the source to the destination.

3.1.3 Authentication Requirements

The Hermes framework can be applied, in principle, to any MANET. We remark that cryptographic primitives should be employed to ensure the security of the trust establishment phase in conjunction with the routing protocol. For example, the exchange of trustworthiness values $T_{i,j}$ should be authenticated and protected by cryptographic primitives. In Section 4.5, we propose an authentication scheme for both data packets and control packets used for trust establishment. We assume that the nodes have already established a set of pairwise keys using a key management protocol [9, 49, 64, 65]. If a secure routing protocol is in place [9–11, 13, 15, 66, 67], the keys established for secure routing can also be used to secure the Hermes scheme.

3.2 Trust Evaluation from First-hand Observations

In order to establish trust, raw data or observations must be accumulated from the network and transformed into numerical trust values. The accumulation of evidence from the observations is discussed in Section 3.4. In this Section, we describe our approach to computing trust given a set of observations obtained from the network. Our approach is based on representing trust in terms of a probability distribution, in particular, the Beta distribution, which is updated using a Bayesian framework. The use of a Bayesian framework was also proposed in [7], but our approach to trust establishment is different in that two separate, but complementary trust metrics are considered: trust and confidence. Further, our approach addresses the issue of incorporating both trust and confidence into trust-based decisions by combining these metrics into a single metric, called *trustworthiness*.

3.2.1 Bayesian Framework

In the Bayesian framework, a random variable R, taking values on the interval $[0,1]$, is associated with a given node. The random variable R represents a notion of trust and is assumed to follow a Beta distribution. A realization of R is taken to be the trust value associated with the node. Since R is assumed to be Beta distributed, trust is represented by the two parameters of the Beta distribution.

The Beta distribution is used because of its reproducibility property under the Bayesian framework. For a given node i, we define a sequence of random variables $R_1, R_2, \cdots$, where R_k characterizes the trust value at the sampling time k. For example, suppose that at time k, N_k network observations have been collected for a given node i. In particular, N_k is the number of packets that have been sent to the node i to be forwarded on to other nodes. Let

M_k be the number of packets actually forwarded by the node, out of the N_k packets that were sent to node i for forwarding at time k. Suppose a prior probability density function (pdf) for R_{k-1}, denoted by $f_{k-1}(r)$ is known. Then the posterior pdf of R_k (given that $N_k = n$ and $M_k = m$) can be obtained from Bayes theorem [68] as follows:

$$f_k(r) = \frac{f_k(M_k = m | r, N_k = n) f_{k-1}(r)}{\int_0^1 f(M_k = m | r, N_k = n) f_{k-1}(r) dr}, \tag{3.1}$$

where $f_k(M_k = m | r, N_k = n)$ is called the likelihood function and has the form of a binomial distribution:

$$f(M_k = m | r, N_k = n) = \binom{n}{m} r^m (1 - r)^{n-m} \tag{3.2}$$

The prior pdf $f_{k-1}(r)$ summarizes what is known about the distribution of R_{k-1}. Under the assumption that the prior pdf $f_{k-1}(r)$ follows a Beta distribution, it can be shown that the posterior pdf $f_k(r)$ also follows a Beta distribution. The Beta distribution with parameters a and b is defined as follows:

$$\text{Beta}(a, b) = \frac{r^{a-1}(1-r)^{b-1}}{B(a, b)} = \frac{r^{a-1}(1-r)^{b-1}}{\int_0^1 r^{a-1}(1-r)^{b-1} dr} \tag{3.3}$$

for $0 \leqslant r \leqslant 1$. In particular, if

$$f_{k-1}(r) \sim \text{Beta}(a_{k-1}, b_{k-1}),$$

then given that $N_k = n_k$ and $M_k = m_k$ we have

$$f_k(r) \sim \text{Beta}(a_{k-1} + m_k, b_{k-1} + n_k - m_k).$$

Therefore, $f_k(r)$ is characterized by the parameters a_k and b_k, defined recursively as follows:

$$a_k = a_{k-1} + m_k \text{ and } b_k = b_{k-1} + n_k - m_k.$$

At the system initiation (at time k = 0), there is no information for the network. Therefore, we assume that R_0 has the uniform distribution over the interval $[0, 1]$, i.e.,

$$f_0(r) \sim U[0, 1] = \text{Beta}(1, 1),$$

which indicates our ignorance about the node's behavior at time 0.

3.2.2 Trust and Confidence Values

We define the trust value, t^k, assigned to a node at time k, to be equal to the mean value $\mu(a_k, b_k)$ of the $\text{Beta}(a_k, b_k)$ distribution corresponding to the pdf $f_k(r)$ as follows:

$$t^k \triangleq \mu(a_k, b_k) = \frac{a_k}{a_k + b_k}, \tag{3.4}$$

for $0 \leqslant \mu \leqslant 1$. We define the confidence value, c^k, associated with the trust value t^k in terms of the standard deviation $\sigma(a_k, b_k)$ corresponding to the pdf $f_k(r)$ as follows:

$$\begin{aligned} c^k &\triangleq 1 - \sqrt{12}\, \sigma(a_k, b_k) \\ &= 1 - \sqrt{\frac{12 a_k b_k}{(a_k + b_k)^2 (a_k + b_k + 1)}} \end{aligned} \tag{3.5}$$

16

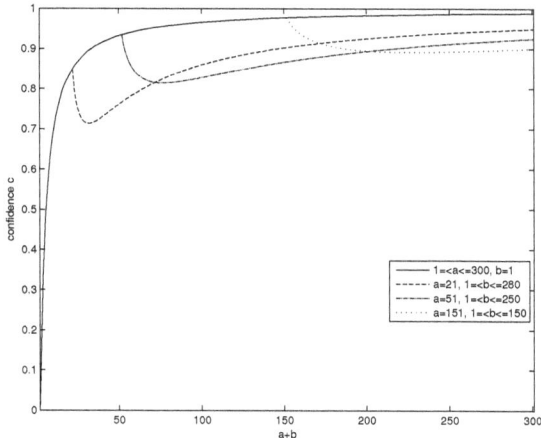

Figure 3.2: Confidence as a function of number of observations.

where $0 \leqslant c^k \leqslant 1$. A value of c^k close to one indicates high confidence in the accuracy of the computed trust value t_k, whereas a value close to zero indicates low confidence.

The definition of confidence value (3.5) captures the statistical dispersion from the mean value of the distribution, which corresponds to the trust value, as defined in (3.4). Note that the closer the kth Beta probability distribution corresponding to $f_k(r)$ approximates a Dirac function, the more confidence is placed in the trust value t^k. A Dirac function indicates absolute certainty. At system initialization time $(k = 0)$, the trust value assigned to each node is given by $t^0 = \mu(1,1) = 0.5$ which indicates our ignorance about the node's behavior. If we take the value 0.5 as the threshold that must be exceeded in order to consider a node to be trusted, then at time 0 a node is considered neither trusted $(0.5 < \mu \leqslant 1)$, nor misbehaving $(0 \leqslant \mu < 0.5)$. The associated confidence value is $c^0 = 0$ according to (3.5). Fig. 3.2 shows how confidence grows when the number of observations grows, for different values of the parameters a and b of the Beta distribution.

3.2.3 Trustworthiness

As discussed in the previous Section, at each time instant k a given node can be characterized by a pair (t^k, c^k). In particular, node i characterizes its trust in node j at time k by the pair $(t_{i,j}^k, c_{i,j}^k)$. However, using a pair of values to describe the opinion of a node for another node, makes comparisons between different nodes difficult. In particular, it is difficult for a given node to decide which of its neighbors is more "trusted" given the corresponding set of (t, c) values. We shall now present a flexible method to transform the pair of values (t, c) into a single value T, which we call *trustworthiness*.

We note that given a pair (t, c) assigned to a node, the greater the value of c, the more

17

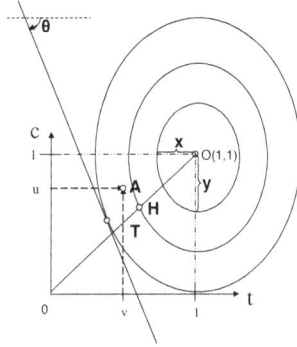

Figure 3.3: Relationship between trustworthiness T and the trust-confidence pair (t, c).

the value of t can be considered as correctly reflecting the trust associated with the node. In this case, the trustworthiness metric should closely reflect the trust value, since the statistical confidence in this value is high. On the other hand, the smaller the value of c, the less the value of t should be considered as valid. If the value of confidence is low, then the reliability of the trust value should correspondingly be low. In this case, the trustworthiness metric should reflect the degree of uncertainty implied by the confidence value. Therefore, for large values of confidence, the trust value t should be weighted more than the confidence value c. Conversely, for small values of confidence, the trust value t should be weighted less than the confidence value c.

Fig. 3.3 shows that the set of (t, c) values lies in the unit square region defined by $0 \le t \le 1$ and $0 \le c \le 1$. For example, the point A corresponds to the pair (u, v). In order to define trustworthiness, each pair (t, c) in the unit square must be mapped into a single value T. There are many ways to define the mapping from (t, c) to T. Fig. 3.3 illustrates the approach we have taken, which is based on considering the family of ellipses centered at the point $(1, 1)$, defined as follows:

$$\frac{(t-1)^2}{x^2} + \frac{(c-1)^2}{y^2} = 1, \tag{3.6}$$

where the pair of values of (x, y) defines the size and shape of the ellipse. The portion (if any) of the (x, y)-ellipse that lies in the unit square determines the set of (t, c) pairs that are mapped to a common value of trustworthiness defined by:

$$T \triangleq 1 - \frac{\sqrt{\frac{(t-1)^2}{x^2} + \frac{(c-1)^2}{y^2}}}{\sqrt{\frac{1}{x^2} + \frac{1}{y^2}}} \tag{3.7}$$

The tangent line of a point (t, c) in the unit square lying on an ellipse with fixed parameters

18

x and y, dictates the relationship between (t, c) and the trustworthiness value T. Let θ denote the angle between the tangent line and the t-axis. The value of θ lies in the interval $[-\pi/2, 0]$ and determines the mapping from (t, c) to T as follows:

- For $\theta = 0$, the value of t is ignored, i.e., $T = c$.

- For $-\pi/4 \leqslant \theta < 0$, the value of c weighs more heavily than the value of t in determining T.

- For $\theta = -\pi/4$ the values of t and c weigh equally in determining T.

- For $-\pi/2 < \theta < -\pi/4$, the value of t weighs more than the value of c .

- For $\theta = -\pi/2$, the value of c is ignored, i.e., $T = t$.

We now consider the impact of the choice of parameters x and y (i.e., the choice of ellipse) on the mapping of (t, c) to T. We will also refer to the x and y parameters as trustworthiness parameters.

- When $x > y$, the angle of the tangent to the ellipse at points (t, c) in the unit square takes values in the interval $(-\pi/4, 0]$ for the majority of the ellipse's points (within the unit square). This implies that that the confidence value has greater weight than the trust value for the majority of points on the ellipse.

- When $x = y = r$, the ellipse becomes a circle of radius r. The tangent line at the point $H = (t_H, c_H)$ in Fig. 3.3 has an angle of $\theta = -\pi/4$. At the point H, the values of t and c have equal weight in determining T, i.e., $T = (t + c)/2$. For all points (t, c) on the ellipse that lie below H (i.e., $c < c_H$), the value of c has a larger weight than the value of t in determining T. Conversely, for all points (t, c) on the ellipse lying above H, the value of t has a larger weight than c in determining T.

- When $x < y$, the angle of the tangent to the ellipse at points (t, c) in the unit square takes values in the interval $[-\pi/2, -\pi/4)$ for the majority of the ellipse's points (within the unit square). This implies that the trust value has greater weight than the confidence value for the majority of points on the ellipse.

Intuitively, the parameters should be chosen such that $x < y$ in order to define an appropriate mapping from (t, c) to T. From Fig. 3.2, we see that after a sufficient number of observations, the value of confidence c grows to a large value. In this range of values of c, the trustworthiness value T should be determined primarily by the value of t. Ultimately, the goal of the Hermes scheme is to distinguish between two classes of nodes: 1) "good" nodes, which forward packets reliably; and 2) "bad" nodes, which do not forward packets reliably, together with nodes for which insufficient statistical evidence is available. As discussed above, the mapping (3.7) provides a concrete rule for combining the notions of trust and confidence into a single metric to allow a node to decide whether another node is "good." The performance of the proposed mapping in terms of convergence, false alarm, and missed detection rates is investigated through computer simulations in Section 3.7.3. We also remark that the choice of (x, y) could also be left up to the node that computes the trustworthiness value of the other node. As will be discussed in Section 3.6, each source node could implement its own policy in determining the trustworthiness of a path by making its own choice of (x, y).

19

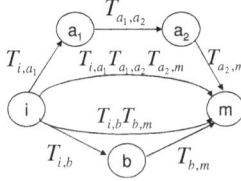

Figure 3.4: Example of opinion calculation for non-neighbors i and m.

3.3 Formulation of Opinions

The concept of trustworthiness as defined in the previous Chapter applies only to two neighbor nodes i and j. Due to the characteristics of the wireless access medium, node i can observe the packet forwarding behavior of node j and then compute the trustworthiness value $T_{i,j}$. However, for the purposes of routing or other network-related decisions, node i may need to form an opinion about an arbitrary node m, which may not be a neighbor of i. Therefore, we generalize the notion of trustworthiness to the concept of *opinion*, which incorporates second-hand trustworthiness values from third-party nodes. The propagation of trustworthiness information to form an opinion is similar to the concept of "recommendations" discussed in [7].

3.3.1 Definition of Opinion

We denote the opinion that node i has for node m by $P_{i,m}$. If node i and m are neighbors, the opinion that i has for m is set equal to the trustworthiness value, $T_{i,m}$, that node i has for m. Recall from Section 3.2 that trustworthiness can be computed from first-hand observation data. If node i and m are not neighbors, neither node can accumulate first-hand information about the other node's packet forwarding behavior. In order for node i to form an opinion about node m, it can make use of the trustworthiness values computed by neighbor nodes within the network.

Assume that i and m are not neighbor nodes. Let $R = \{i = a_0, a_1, a_2, \cdots, a_{n-1}, a_n = m\}$, where $n \geq 2$, denote a path from node i to node m. We can extend the concept of trustworthiness between two neighbor nodes to trustworthiness along a path from one node to another node. More precisely, we define the trustworthiness of path R as follows:

$$j^* \triangleq \min\{\arg\min_{1 \leq j \leq n-2}\{T_{a_j,a_{j+1}} < T_{def}\}, n-1\}. \tag{3.8}$$

$$T_R \triangleq T_{a_0,a_1} \cdot \prod_{j=1}^{j^*} T_{a_j,a_{j+1}} \cdot (T_{def})^{n-j^*-1}. \tag{3.9}$$

If $T_{a_j,a_{j+1}} \geq T_{def}$, node a_j makes use of the trustworthiness values propagated by node a_{j+1}.

Otherwise, node a_j simply sets the trustworthiness values of the downstream nodes to the value T_{def}. Thus, the Hermes scheme correlates the reliability of a node for trust propagation with its trustworthiness for packet forwarding. The impact of doing this on the security properties of Hermes is discussed in Section 3.5.4.

As an example of computing the trustworthiness of a path R acccording to (3.9), suppose that node i receives trustworthiness values $T_{i,a_1}, T_{a_1,a_2}, \cdots, T_{a_{n-2}}$, all of which exceed the value T_{def} and that $T_{a_{n-2},a_{n-1}} < T_{def}$. In this case, node i ignores the trustworthiness value T_{a_{n-1},a_n} that node a_{n-1} has sent. Then, the trustworthiness value T_R of the path R is computed as

$$T_R = T_{i,a_1} \cdot T_{a_1,a_2} \cdot \cdots \cdot T_{a_{n-2},a_{n-1}} \cdot (T_{def})^1.$$

Let $\mathcal{R}_{i,m}$ denote the set of paths from node i to node m. If the set $\mathcal{R}_{i,m}$ is empty, we define $P_{i,m} = T_{def}$, where T_{def} is the default trustworthiness value assigned to a node, when its assigned trust and confidence values are $t^0 = 0.5$ and $c^0 = 0$, respectively. That is, $T_{def} \triangleq T(0.5, 0)$. Otherwise, we define the opinion that node i has for node m by

$$P_{i,m} \triangleq \max_{R \in \mathcal{R}_{i,m}} T_R. \tag{3.10}$$

Since node i is not a neighbor of node a_j, it has to rely on the trustworthiness values computed by other nodes in order to form its own opinion about a_j. That is, node i must use *second-hand information* to form an opinion about m, as can be seen from (3.10). An example of opinion calculation for non-neighbors i and m is illustrated in Fig. 3.4. In this example, all trustworthiness values are greater than T_{def}, i.e., $T_{i,a_1}, T_{a_1,a_2}, T_{a_2,m}, T_{i,b}, T_{b,m} \geq T_{def}$. We now provide a definition for the opinion that any node i has for another node m as follows:

$$P_{i,m} = \begin{cases} T_{i,m}, & i \text{ and } m \text{ are neighbors,} \\ \max_{R \in \mathcal{R}_{i,m}} T_R, & \mathcal{R}_{i,m} \neq \emptyset, \\ T_{def}, & \text{otherwise.} \end{cases} \tag{3.11}$$

3.3.2 Opinions from Second-hand Trustworthiness

When node i and m are not neighbors, the value of $P_{i,m}$ is obtained by computing the maximum value of the trustworthiness values with respect to each path from i to m. This computation can be carried out using a shortest path algorithm by defining a suitable set of edge weights for the network. Define the weight of the link from a node a to a neighbor node b as follows:

$$w_{a,b} \triangleq -\log(T_{a,b}), \tag{3.12}$$

where $T_{a,b}$ is the trustworthiness value that node a has for node b, computed using first-hand information. Note that since $T_{a,b} \in (0,1)$, the value of $w_{a,b}$ must be nonnegative.

Proposition 1. *If i and m are not neighbors, and at least one path exists between them, then*

$$P_{i,m} = \exp(-d_{i,m}), \tag{3.13}$$

where $d_{i,m}$ is the length of the shortest path from i to m.

Proof. The weight of a path $R = \{i, a_1, \cdots, a_n, m\}$ in the network is then defined as the sum of the weights of the edges in the path:

$$w_R \triangleq w_{i,a_1} + w_{a_1,a_2} + \cdots + w_{a_{n-1},a_n} + w_{a_n,m}. \tag{3.14}$$

21

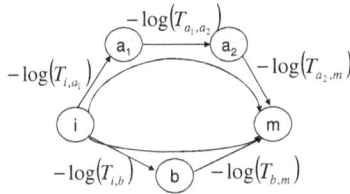

Figure 3.5: Opinion computation as a shortest path problem.

The length of the shortest path from node a to b is then given by

$$d_{i,m} \triangleq \min_{R \in \mathcal{R}_{i,m}} w_R. \qquad (3.15)$$

Now it can easily be verified that

$$P_{i,m} = \exp(-d_{i,m}).$$

$\square$

The mapping of the opinion computation to a shortest path problem is illustrated in Fig. 3.5. In MANETs, the computation can be performed in a distributed manner using a Bellman-Ford type algorithm [69]. Furthermore, the computation can be "piggybacked" relatively easily onto distance vector routing protocols such as AODV [70].

3.3.3 Propagation of Trustworthiness

The trustworthiness values can be carried by the control packets of the routing algorithm in order to avoid additional communication overhead. Therefore, the propagation of the trustworthiness values depends on the routing algorithm in place. Section 3.6 discusses how the trustworthiness values are exchanged in the signaling phases of the DSR [4] and AODV [5] routing algorithms, which are representative examples of source routing and distance vector routing algorithms, respectively.

3.4   Accumulation of Empirical Evidence

In the Hermes framework, a given node collects first-hand observation data with respect to each of its neighbors. The first-hand observations are used to compute trust and confidence values, which are in turn mapped to trustworthiness values. Recommendations are used by the given node to form opinions about non-neighbor nodes. The accumulation of observation data depends on the type of routing algorithm in place. We discuss how observation data can be collected in the case of source routing and distance vector routing. We also propose windowing mechanisms to systematically expire old observation data in order to maintain the responsiveness of the system.

3.4.1 Physical and MAC Layer Assumptions

At the physical layer, we assume that the nodes are equipped with omni-directional antennas and that they transmit at a constant power level, i.e., no power control is used. We shall assume that acknowledgements (ACKs) at the MAC layer are used to verify the successful reception of a packet through the wireless channel. The MAC layer ACKs are sent by the destination hop to notify the source hop that the sent packet has been received. When a MAC layer ACK is not received, the source hop has to resend the unacknowledged packet.

3.4.2 Accounting for Malicious Behavior

Forwarding Packets

A given node X on a path forwards packets to the next or downstream node Y. Suppose that node Z is the next node after node Y on the path. Due to the broadcast nature of the wireless medium, node X could determine, for each packet it forwards to node Y, whether node Y correctly forwards the packet on to node Z. In order to do this, the MAC layer of a node must be modified to forward all received frames to the network layer. In this case, the node is said to be operating in *promiscuous* mode. Thus, node X should process, at the network layer, any packet received at the MAC layer from the wireless interface, whether or not node X is the MAC-level destination of the packet.

In our proposed scheme for accumulating observation data, which has some similarities with the watchdog solution discussed in [71], each node operates in promiscuous mode. When a given node on a source route, say node X, forwards a packet p to the next hop, say node Y, it increments a counter, $C_{X,Y}$, by one and starts a timer. The timeout value should be larger than the round-trip delay between node X and Y. If node X sees a packet from node Y that matches the packet p within the timeout period, then node X is assured that node Y correctly forwarded packet p to the next hop (i.e., node Z) and increments a counter, $F_{X,Y}$. Otherwise, if the timeout period expires, node X assumes that node Y did not forward packet p on to node Z. We point out that the penultimate node in the route, i.e., the node immediately upstream from the destination node D, does not expect node D to forward packets and hence does not follow this procedure.

Note that the set of active traffic flows traversing node X and the neighbor set of node X change over time. Therefore, node X can potentially accumulate packet delivery statistics for every other node in the network. The set of values $C_{X,y}$ and $F_{X,y}$ for all other nodes y in the network forms a table of *Packet Delivery Statistics*, which can be used to compute the first-hand trust and confidence values $t_{X,y}$ and $c_{X,y}$, respectively, according to the Bayesian framework discussed in Section 3.2.1. The pair $(t_{X,y}, c_{X,y})$ can then be mapped to a trust-worthiness value $T_{X,y}$, as discussed in Section 3.2.3.

In our proposed scheme for accumulating packet observation data, nodes maintain packet counters corresponding only to packets they have actually forwarded. The scenario of Fig. 3.6 illustrates a potential problem that may arise if a node attempts to accumulate information on packets that it did not forward. Here, node I wishes to send a packet to node F along the path $\{I, E, F\}$. Suppose that node E behaves correctly and forwards the packet to node F. In promiscuous mode, node A will hear the packet that node E forwards (since node E is in node A's radio reception range), but will not know how many packets node E received for forwarding, since node I is not in node A's radio reception range. Therefore, we limit our

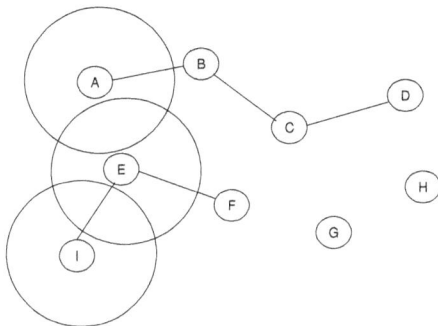

Figure 3.6: Example topology showing active flows on paths $\{I, E, F\}$ and $\{A, B, C, D\}$.

scheme to gathering statistics only for packets that a node has forwarded itself, to ensure that valid information is recorded. On a route of n nodes (including the source and destination nodes), the first $n - 2$ nodes accumulate evidence for their downstream nodes.

Misrouting Packets

Node X could determine, for each packet it forwards to node Y, whether node Y correctly forwards the packet on to node Z. In order to do this, node X must operate in promiscuous mode. When node X forwards a packet p to the next hop node Y, it increments the counter, $C_{X,Y}$, by one and starts a timer. The timeout value should be larger than the round-trip delay between node X and Y. If node X sees a packet from node Y that matches the packet p sent to node Z within the timeout period, node X is assured that node Y correctly forwarded packet p to the next hop and increments the counter $F_{X,Y}$. Otherwise, if the timeout period expires or if the packet was not forwarded to node Z, node X does not increment the counter $F_{X,Y}$. We point out that the penultimate node in the route, i.e., the node immediately upstream from the destination node D, does not expect the destination node D to forward packets. Hence, the penultimate node does not follow this procedure.

Injecting Packets

A node injects packets when it sends new packets into the network and attributes them to a flow of another node. When a secure routing algorithm is implemented, a node cannot inject packets unless it has access to the secret key of the source node. Thus, a node cannot drop the legitimate packets and inject new packets in order to let its upstream node believe that it forwarded the packets it received for forwarding. In case the secret key of a node is compromised, packets can be injected by that node. However, this issue is beyond the scope of this book.

3.4.3 Routing Protocol Considerations

In source routing protocols, e.g., DSR [72], each datagram at the network layer contains the list of nodes in the entire route from the source to the destination. Therefore, a node X, operating in promiscuous mode, can recognize whether its downstream node Y correctly forwards a packet p on to its downstream node Z. Node X operates in promiscuous mode. When node X sees a packet sent from node Y within the timeout period, node X checks the packet by looking at the source route listed in the datagram's header to see whether the packet matches packet p, and the destination field listed in the frame's header to see whether the packet is sent to the correct next hop. In case the packet matches packet p, and is sent to node Z, node X is assured that node Y correctly forwarded packet p to the next hop.

In distance vector routing algorithms such as AODV [70], the header of a data packet contains information about the next hop and the number of remaining hops to the destination. Upon receiving a data packet, a node overwrites the next hop field and decreases the number of hops left to the destination by one. The observation scheme presented earlier for source routing does not work for distance vector routing because a node that sends a packet to its downstream node for forwarding cannot determine whether the packet will indeed be forwarded, as the upstream node's identity does not appear in the new header of the packet.

A simple and efficient solution is to employ sequence numbers at the network layer to identify each data packet during the data forwarding phase. By checking the sequence number, a given node X can then verify whether its downstream neighbor node Y correctly forwarded a given packet p that was sent earlier by X. Nonetheless, node X does not know which is the downstream node of node Y. Therefore, node X can only verify whether node Y correctly forwarded a given packet p, but does not know whether node Y misroutes the packet p to node V. In this case, node V is responsible for forwarding the packet towards its destination. Thus, the packet might traverse a longer route to the destination node.

We note that a malicious node could misroute packets to a colluding node that drops the packets. One approach to avoid such an attack is to extend AODV routing tables to carry more than the next hop information. A full development of this approach is beyond the scope of this book.

3.4.4 Observation and Averaging Windows

Observation Window

We define an observation window over which a given node i collects first-hand observation data from its neighbor node j. At the end of the kth observation window, denoted by W_k, the trustworthiness value $T_{i,j}^k$, of node i for node j is calculated using the observations from W_k. We assume that each observation window is of length τ. Given the trustworthiness values $T_{i,j}^k$, the set of opinion values corresponding to window W_k, i.e., $\{P_{i,m}^k\}$ for any node m, can be computed. As indicated in Fig. 3.7, the computation of $P_{i,m}^k$ is assumed to take an additional τ_P time units after window W_k ends, during the first part of window W_{k+1}.

Averaging Window

We propose a sliding windowing mechanism to systematically expire old observation data in order to improve the accuracy of the opinion metric and maintain the responsiveness of

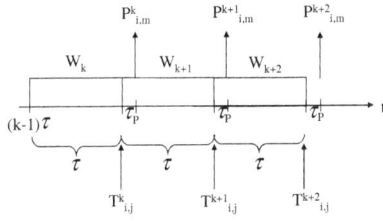

Figure 3.7: Sequence of observation windows.

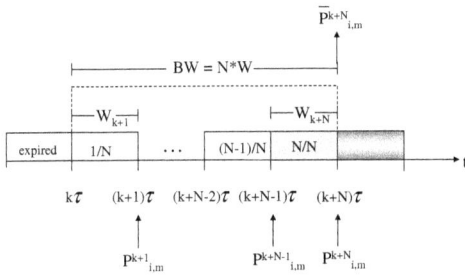

Figure 3.8: Averaging window.

the system. The sliding averaging window BW_k consists of the N most recent observation windows, i.e.,

$$BW_k = \{W_{k-N+1}, W_{k-N+2}, \cdots, W_{k-1}, W_k\}. \tag{3.16}$$

The length of BW_k is $N\tau$ time units. During the averaging window BW_k, N opinion values are computed for each pair of nodes i and m (see Fig. 3.8):

$$P_{i,m}^{k-N+1}, P_{i,m}^{k-N+2}, \cdots, P_{i,m}^{k-1}, P_{i,m}^k, \tag{3.17}$$

which correspond to the N observation windows contained in BW_k. We calculate a weighted average of the N opinion values computed during the window BW_k to obtain an *averaged opinion value*, $\bar{P}_{i,m}^k$. By applying a simple linear weighted averaging scheme, we define the averaged opinion at time k that node i has for node m as follows:

$$\bar{P}_{i,m}^k \triangleq \frac{2}{N(N+1)} \sum_{l=1}^{N} l P_{i,m}^{k-N+l}. \tag{3.18}$$

We remark that other averaging schemes, e.g., exponential averaging windows, may also be used to define the averaged opinion. The use of averaging improves the stability of the opinion metric, since past information is taken into account.

3.5 Attacker Model and Security Properties of Hermes

3.5.1 Attacker Models for MANETs

Most of the attacker models for MANETs discussed in the literature are presented in the context of secure routing [12, 14, 73]. Attacks against routing protocols can be classified into passive and active attacks. The primary example of a passive attack is eavesdropping of routing control packets. Examples of active attacks against a routing protocol include: black or grey holes, creation of routing loops, wormhole attacks, gratuitous detour attacks (i.e., making a route appear shorter than it is), corrupting packets, fabricating packets, replaying packets, reordering packets, misrouting packets, and impersonating another user.

Many of the aforementioned attacks can be addressed using cryptographic mechanisms. For example, message authentication codes (MACs) can be applied to protect the integrity of control packets. The use of cryptography can also prevent impersonation attacks. Other attacks on routing can be mitigated using sequence numbers. For instance, the numbering of route request packets can be used to avoid the creation of routing loops. The route request and route reply phases of a routing protocol can be used to mitigate attacks such as black and grey holes and packet misrouting. Other mechanisms, such as packet leashes [74] have been proposed to address attacks such as wormhole attacks.

3.5.2 Attacks Addressed by Hermes

The Hermes framework is intended to avoid a class of attacks on packet delivery in MANETs during the data transmission phase rather than the route discovery phase. The attacks on Hermes are considered to be committed by "insider" nodes who have succeeded in becoming part of active routes in the network. Such nodes are owners of valid cryptographic keys or key materials and are capable of authenticating themselves as authorized users of the network. Furthermore, these nodes have successfully passed the route discovery phase of a secure routing protocol.

As mentioned earlier, the integrity of message exchanges involved in the Hermes protocol should be protected by cryptographic primitives such as those used in secure routing protocols. The authentication of the packets exchanged during trust establishment is discussed in Section 4.5. Here, we focus our attention on insider attacks on packet forwarding and the propagation of trust information. The main attacks on packet forwarding to be considered in the attacker model include dropping, misrouting, and replaying data packets. As in secure routing protocols, sequence numbers can be used in conjunction with Hermes to avoid replay attacks. The main focus of the Hermes scheme lies in detecting packet dropping and misrouting attacks.

We remark that packet forwarding attacks can be launched even when a secure routing protocol is in place. A secure routing protocol aims to establish a route from a source node to a destination node containing only authorized or insider nodes. Once a route is established, nodes on the path are expected to forward packets correctly to the next hop. However, during the data transmission phase an insider node may consistently drop, misroute, or replay packets. The Hermes scheme attempts to identify such misbehaviors in terms of the trustworthiness and opinion metrics, but does not purport to distinguish between malicious or non-malicious misbehaviors. Non-malicious packet forwarding misbehavior may be due to such phenomena as network congestion, node mobility, or node malfunction. Nevertheless, we consider such behavior to be untrustworthy.

To compute the opinion metric for non-neighbor nodes, the Hermes scheme relies on the exchange of trustworthiness information among nodes. Thus, an obvious attack on the Hermes scheme would be for nodes to propagate false trustworthiness information. In the basic Hermes framework, the trust that a node X has in the trustworthiness information propagated by another node Y is simply correlated with the trust that node X has for node Y with respect to packet forwarding. Consequently, an attack that cannot be resolved by the Hermes scheme is one in which node Y propagates false trust information to node X, yet forwards packet correctly. An extension to Hermes, which avoids such an attack is discussed in Chapter 4.

3.5.3 Probabilistic Attacker Model

The attack space covered by the Hermes scheme can be described more formally in terms of a probabilistic attacker model. The attacker model consists of two types of attacks: 1) incorrect data packet forwarding; 2) incorrect propagation of trust information. Note that we do not distinguish among the various types of data packet forwarding misbehaviors, i.e., packet dropping, misrouting, and replay attacks. Incorrect trust propagation refers to a node which propagates a trustworthiness value that is different from the value that it would compute if it were following the Hermes scheme. Thus, a node may propagate a trustworthiness value that is higher or lower than the value that a Hermes-compliant node would compute.

Let $\mathcal{N}$ denote the set of all nodes in the network. A network attack scenario, in steady-state, is specified by characterizing, for each node $i \in \mathcal{N}$, the probability B_f^i that the node performs incorrect packet forwarding and the probability B_t^i that the node performs incorrect trust propagation, where $0 \leq B_f^i, B_t^i \leq 1$. More precisely, the network attack scenario can be represented by a set of three-tuples,

$$\mathcal{S} = \{(i, B_f^i, B_t^i) : i \in \mathcal{N}\}. \tag{3.19}$$

Let η_f and η_t denote, respectively, thresholds on the degrees of packet forwarding and trust propagation misbehaviors that can be tolerated in the network (e.g., one could set $\eta_f = \eta_t = 0.5$). A given node i can then be identified as an *attacker* if $B_f^i > \eta_f$ or $B_t^i > \eta_t$ (or both).

In the context of the Hermes scheme, which is primarily concerned with packet forwarding behavior, a *good* node is one for which $B_f \leq \eta_f$, whereas a bad node is one for which $B_f > \eta_f$. The Hermes scheme aims to identify the set of probabilities $\{B_f^i\}$ to a sufficient degree of accuracy to distinguish between good and bad nodes, based on both first-hand information from direct observations of packet forwarding behavior and second-hand information via recommendations obtained from other nodes. The probabilistic attacker model does not preclude the possibility that nodes may collude with one another. However, the Hermes scheme does not seek to identify collusions per se. Rather, the Hermes scheme is able to characterize the *effect* of a colluding attack as represented by an attack scenario (3.19).

3.5.4   Security Properties of Hermes

The attacker model presented above is simple, but sufficient to characterize the main security properties of the Hermes scheme. Under Hermes, the opinion metrics $P_{i,m}$ should closely approximate the underlying attack scenario under steady-state conditions. That is, in steady-state we should have

$$P_{i,m} \approx 1 - B_f^m, \quad \text{for all } i \in \mathcal{N}. \tag{3.20}$$

For the Hermes framework to correctly distinguish the good nodes from the bad nodes, it is sufficient that

$$P_{i,m} \geq T_{def} \text{ if and only if } B_f^m < \eta_f, \quad \text{for all } i \in \mathcal{N} \tag{3.21}$$

hold in steady-state. The simulation results presented in Section 3.7.3 provide validation of the steady-state properties (3.20) and (3.21).

We point out that the basic Hermes framework presented in this Chapter assumes that the trust propagation behavior of a given node is correlated with its packet forwarding behavior in the following sense. According to (3.9), the trustworthiness values propagated by a node i to a neighbor node j are effectively ignored by node j if the trustworthiness value $T_{j,i}$ is less than the default value T_{def}. Consequently, in terms of the probabilistic attacker model, the Hermes scheme can correctly infer packet forwarding behavior under the following assumption:

$$B_f^i \geq B_t^i, \quad \text{for all } i \in \mathcal{N} . \tag{3.22}$$

In other words, the probability with which a node commits a packet forwarding misbehavior must not exceed the probability with which it commits a trust propagation misbehavior. To relax assumption (3.22), the Hermes scheme can be extended to maintain an opinion metric with respect to trust propagation, as will be discussed in Chapter 4.

Under the probabilistic attacker model and the assumption (3.22), the Hermes scheme is able to distinguish the good nodes from the bad nodes in a network scenario with high accuracy, as demonstrated through the simulation results presented in Section 3.7.3. In steady-state, the performance of the Hermes scheme with respect to a compliant node does not depend on the set of probabilities $\{B_f^i : i \in \mathcal{N}\}$ or the locations of the nodes. However, the convergence of the scheme is a function of the availability of first-hand observation data, which depends on the distribution and traffic volume of flows in the network. The probabilistic

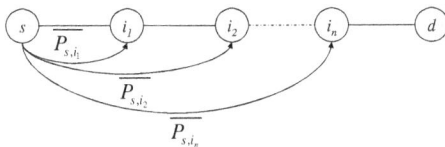

Figure 3.9: Calculation of routing opinion.

attacker model only characterizes steady-state behavior. To accommodate dynamic changes in the network attack in practice, the proper use of windowing as discussed in Section 3.4.4 is necessary to maintain the responsiveness of the Hermes scheme.

3.6 Trust-Aware Routing

In this Section, we discuss the application of the Hermes trust establishment framework to improve the reliability of packet forwarding in MANET routing protocols in the presence of malicious nodes. First, we define the concept of "routing opinion," which is used to select among a set of alternative routes from a source to a destination node. Then we briefly discuss how the Hermes framework can be incorporated in the well-known MANET routing protocols DSR and AODV, respectively.

3.6.1 Definition of Routing Opinion

Given a source node s, a destination node d, and a path $R = \{s, i_1, \cdots, i_n, d\}$ from s to d, we define the "routing opinion" that node s has for the route R as follows:

$$V_R \triangleq (\bar{P}_{s,i_1} \cdot \bar{P}_{s,i_2} \cdot \cdots \cdot \bar{P}_{s,i_{n-1}} \cdot \bar{P}_{s,i_n})^{1/n}. \tag{3.23}$$

This definition is illustrated in Fig. 3.9. According to (3.23), the routing opinion of s along route R is a function of the product of the (averaged) opinions that node s has for each node on the path R, except for the destination node d. The reason that $\bar{P}_{s,d}$ is not included in the product is that when node s chooses to communicate with node d, it implicitly trusts node d. The selection of a route entails a choice of intermediate nodes, not including node d, that lie on a path to d. In the definition (3.23) of routing opinion, the exponent $1/n$ is included in order to avoid excessively penalizing longer routes.

3.6.2 Route Selection

Given a source node s, a destination node d, a path $R = \{s = a_0, a_1, a_2, \cdots, a_{n-1}, a_n = d\}$, where $n \geq 2$, from s to d, link l of route R, $l \in R$, trustworthiness T of link l, and the set of paths $\mathcal{R}_{s,d}$ from node s to node d, we define the route R^* on which node s chooses to send its data packets to destination node d as follows:

$$T^* \triangleq \max_{R \in \mathcal{R}_{i,m}} \min_{l \in R} T_l. \tag{3.24}$$

$$\mathcal{C}_{i,m} = \{ R \in \mathcal{R}_{i,m} : \min_{l \in R} T_l = T^* \} \tag{3.25}$$

$$R^* \triangleq \arg \max_{R \in \mathcal{C}_{i,m}} V_R \tag{3.26}$$

According to equation (3.26), node s chooses to send its data packets to destination node d on the route of maximum routing opinion, which is chosen among the routes of the maximum of the minimum link trustworthiness. The rationale for this is that any intermediate link on a route can be point of failure. Finally, the route of maximum routing opinion, among the routes of the maximum of the minimum link trustworthiness, is chosen by source node s to send its data packets to destination nodes d. Alternatively, the routing opinion metric could be used to choose the route probabilistically, i.e., a route would be chosen with probability proportional to the routing opinion. Such a randomized routing scheme would improve the performance of the Hermes scheme, as the flows would traverse a more diverse set of nodes, providing a richer set of first-hand observation data for computing the trust metrics. The competitive adaptive routing scheme proposed by Awerbuch et al. [75], employs a type of randomized routing except that the routing decisions are performed on a packet-by-packet basis rather than a per-flow basis.

3.6.3 Trust-aware DSR

DSR is a reactive ad hoc routing protocol based on source routing [72]. To incorporate the Hermes framework into DSR, each node computes trustworthiness values with respect to each of its neighbors. The routing opinion V_R that a node i has for a route R to node m is only computed when i receives the route reply (RREP) message sent by m in response to the route request (RREQ) message sent from node i to node m. Corresponding to each RREQ message, the destination sends a RREP message. As the RREP message propagates to the source i along a path R, it accumulates the link weights defined in equation (3.12) maintained by the nodes along the path. The source node i then determines its routing opinion, V_R, for route R to node m using equation (3.23). As specified by DSR, each node stores multiple paths to each destination. Our scheme requires the addition of a field indicating the trustworthiness of each route. Node i chooses to send its data packets to destination node m on the route R^* defined by equation (3.26) (see Section 3.6.2).

Recall from equation (3.23) that in order to calculate the routing opinion along a path from node i to node m, the opinion values for the intermediate hops are required. The computation of these opinion values is defined in equation (3.13). According to equation (3.23), the averaged opinion values, $\bar{P}_{i,m}$, should be used to compute the routing opinion, rather than the current opinion values, $P_{i,m}$. Since the source node can compute the opinion values $P_{i,m}$ only when it receives a RREP message, it is problematic for the source node to compute the averaged opinion values. An alternative approach is for each of the nodes in the network to compute *averaged* trustworthiness values, $\bar{T}_{i,j}$, which are averaged over the past N observation windows, similarly, to the averaged opinion values. Then the averaged opinion value could be computed according to the following modified definition (cf. (3.10)):

$$\bar{P}_{i,m}^k \triangleq \max_{R \in \mathcal{R}_{i,m}} \bar{T}_R, \tag{3.27}$$

where (cf. (3.9))

$$T_R \triangleq \bar{T}_{a_0,a_1} \cdot \prod_{j=1}^{j^*} \bar{T}_{a_j,a_{j+1}} \cdot (T_{def})^{n-j^*-1}, \tag{3.28}$$

and (cf. (3.18))

$$\bar{T}_{i,j}^k \triangleq \frac{2}{N(N+1)} \sum_{l=1}^{N} l T_{i,j}^{k-N+l}. \tag{3.29}$$

Each source node can implement its own policy in determining the routing opinion of a path by making its own choice of (x, y) trustworthiness parameters (see Section 3.2). The trustworthiness parameters can be piggybacked in the RREQ messages. Corresponding to each RREQ, the destination sends a RREP containing the trustworthiness parameters send by the source node in the RREQ. Each intermediate node should use the given trustworthiness parameters in computing the averaged trustworthiness values $\bar{T}_{i,j}$ of their downstream node.

3.6.4 Trust-aware AODV

AODV is a reactive ad hoc routing protocol based on distance vector routing [70]. As discussed in Section 3.3.2, the averaged trustworthiness values $\bar{T}_{i,j}$ can be propagated to calculate the averaged opinion values by means of a Bellman-Ford or distance-vector type algorithm. Since AODV is based on distance-vector routing, the propagation of the averaged trustworthiness values $\bar{T}_{i,j}$ can easily be incorporated into AODV via the route reply (RREP) messages. The averaged trustworthiness values $\bar{T}_{i,j}$ are computed as defined by equation (3.29). Then, each node i can compute its opinion $\bar{P}_{i,m}$ for every other node m in the network. The averaged opinion values $\bar{P}_{i,m}$ are calculated as defined by equation (3.27). The source node i determines its routing opinion, V_R, for route R to node m using equation (3.23). Node i chooses to send its data packets to destination node m on the route R^* defined by equation (3.26) (see Section 3.6.2).

As specified by AODV, each node maintains a routing table with the next hop and the hop count to each destination. Our scheme requires the addition of a third field indicating the trustworthiness of each route. When AODV is implemented, the destination nodes respond to the first RREP received, unless one arrives along a better path. Hence, more than one RREP may reach the source node, which can calculate the trust value of the returned routes and choose to route the packets through the most trusted one. AODV specifies that each source node stores one path for each destination. AODV could be extended so that source nodes store all the returned paths, each associated with its trust value. As discussed in the case of DSR, each source node can implement its own policy in determining the routing opinion of a path by making its own choice of (x, y) trustworthiness parameters. The trustworthiness parameters can be piggybacked in the RREQ messages.

3.7 Performance Evaluation

In this section we evaluate the performance of Hermes. We first discuss the convergence and accuracy of the scheme. Then, we evaluate the accuracy of Hermes by presenting some representative results from our simulation experiments.

3.7.1 Convergence of Hermes

Hermes converges to the correct value of the trustworthiness metric T, when the confidence c ($c \in [0, 1]$) associated with the trustworthiness metric T reaches one.

As discussed in Section 3.2.2, the confidence value, c, associated with the trust value t is defined in (3.5). If A denotes the total number of packets forwarded *correctly* (not dropped or misrouted) from the downstream nodes, whereas M denotes the total number of packets sent for forwarding by the downstream nodes, from (3.5), confidence c is also given by:

$$c = 1 - \sqrt{\frac{12(A/M)(1 - A/M)}{M + 1}} = 1 - \varepsilon, \tag{3.30}$$

where

$$\varepsilon = \sqrt{\frac{12p(1 - p)}{M + 1}},$$

with $p = A/M$. It is not hard to show that

$$\varepsilon \leq \sqrt{\frac{12(1/4)}{M + 1}} = \sqrt{\frac{3}{M + 1}} = O\left(\frac{1}{\sqrt{M}}\right). \tag{3.31}$$

Hence, the confidence metric converges to one at a rate of $O(1/\sqrt{M})$.

3.7.2 Accuracy of Hermes

The accuracy of Hermes can be measured by the relationship between trustworthiness T and trust t, 1) when confidence c equals to one, i.e., when Hermes converges to the correct trustworthiness values T and 2) when confidence c equals to zero, i.e., at the initialization time. In principle, when confidence $c = 1$, the trustworthiness value T should approximate trust value t. The trustworthiness metric T was defined in Section 3.2.3 as a function of trust t, confidence c and trustworthiness parameters x and y (see (3.7)). Let us now introduce the parameter r, where $r = x/y$. Then, the trustworthiness associated with a pair (t, c) can also be defined as

$$T(t, c) \triangleq 1 - \frac{\sqrt{(t - 1)^2 + r^2(c - 1)^2}}{\sqrt{1 + r^2}}, \tag{3.32}$$

where r determines the relative importance of the trust value t vs. the confidence value c.

Here, we calculate the optimal trustworthiness parameters, i.e., parameter values that satisfy the conditions: $c = 1 \Rightarrow T = t$ and $c = 0 \Rightarrow T \neq t$. When $c = 1$,

$$T = 1 - \frac{1}{\sqrt{1 + r^2}}(1 - t)$$

We shall choose r to minimize $f(r) = (T - t)^2 = [(1 - \frac{1}{\sqrt{1+r^2}})(1 - t)]^2$. The first derivative,

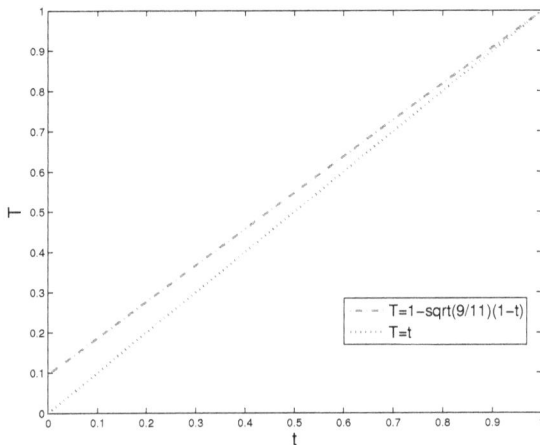

Figure 3.10: Hermes accuracy.

$f'(r)$, is a function of both t and r, which means that the optimal value of the parameter r depends on the value of trust t.

We choose $x = \sqrt{2}$ and $y = \sqrt{9}$ (or $r = \sqrt{2/9}$) based on simulations results presented in Section 3.7.3. Thus, for $c = 1$, the trustworthiness T defines the accuracy of Hermes as follows:

$$T = 1 - \sqrt{\frac{9}{11}}(1 - t) \tag{3.33}$$

Figure 3.10 graphs (3.33) and $T = t$. As expected, the accuracy of Hermes depends on the value of trust t. The maximum error e of Hermes, which occurs when $t = 0$, can be derived from (3.33) and equals to $e = 0.0955$. In general, the error in Hermes' accuracy depends on trust t and is $e \leq 0.0955$. We remark that this error is introduced, because we require that when confidence c is zero, trustworthiness T should not equal to trust t. At system initialization time, the trustworthiness value should capture the notion that there is no confidence for the assumed value of trust $t = 0.5$.

3.7.3 Performance Results

The Hermes scheme was implemented and evaluated in Matlab. We present three simulation scenarios. The network topology shown in Fig. 3.11 is used for the simulations. Fourteen wireless links are formed among ten nodes that are randomly placed in a 1000 m by 400 m area. The wireless radio transmission range of the nodes is set to 250 m. We remark that the simulation scenarios considered in this Chapter do not take into account the effect of mobility. Mobility will generally *improve* the performance of the Hermes scheme, since nodes will have more opportunities to gain first-hand information on all other nodes in the network. Larger

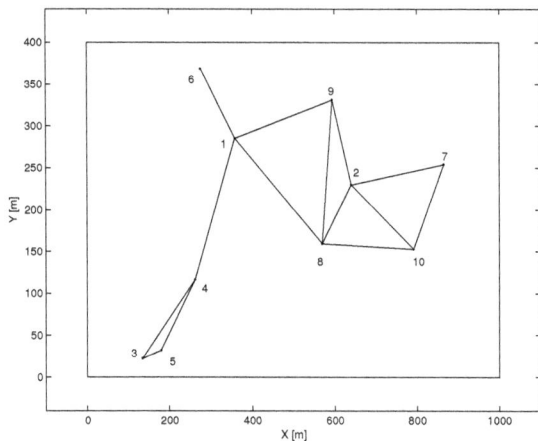

Figure 3.11: Network topology used in simulation experiments.

simulation scenarios which take into account mobility are presented in Chapter 4.

Trust, Confidence, and Trustworthiness

In the first simulation scenario, one traffic flow is established in the network from node 5 to node 7, along the path $\{5, 4, 1, 8, 2, 7\}$. Intermediate nodes 4, 1 and 8 forward 90% of the packets that they should be forwarding, whereas node 2 forwards only 20% of the packets received for forwarding. Node 5 sends 20 data packets during each observation window W. In terms of the probabilistic attacker model discussed in Section 3.5.3, the network scenario can be specified as follows:

$$B_f^i = \begin{cases} 0.1, & \text{for } i = 1, 4, 8 \\ 0.8, & \text{for } i = 2 \\ 0, & \text{otherwise.} \end{cases} \quad \text{and} \quad B_t^i = 0 \text{ for all } i \in \mathcal{N}, \qquad (3.34)$$

where $\mathcal{N} = \{1, \cdots, 10\}$ is the set of all nodes in the network. We remark that the simulation results are the same for any set of $\{B_t^i\}$ satisfying assumption (3.22), i.e., $B_t^i \leq B_f^i$ for all $i \in \mathcal{N}$.

Fig. 3.12 shows the trust and confidence values, $(t, c)_{5,4}$, that node 5 places on node 4 after 0, 1, 3, 10, and 30 windows[1], based on the direct observations of node 5. We note that node 5 forms a correct opinion about node 4, i.e., $(t, c) = (0.85, 0.75)$, even after single round. Observe that the more observations node 5 makes for node 4, the more confident node 5 becomes about the trust value it assigns to node 4.

---

[1] We shall also refer to an observation window as a "round."

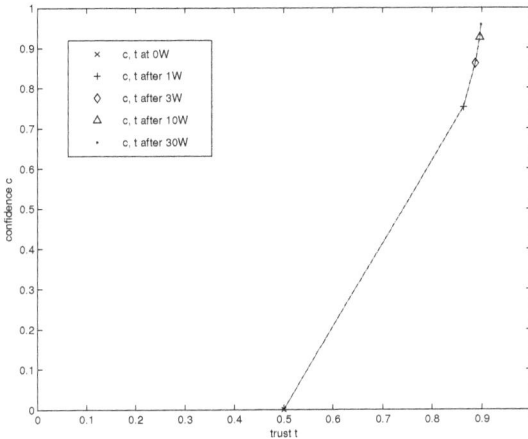

Figure 3.12: $(t, c)_{54}$: (trust, confidence) node 5 places on node 4 after 0, 1, 3, 10 and 30 windows (W).

Figs. 3.13, 3.14, and 3.15 show the opinion values over time (i.e., windows) that node 5 places on node 4, 2 and 3, respectively, for different trustworthiness parameters, x and y. Node 4 is a "good" node, since it forwards 90% of the packets that should be forwarded. Node 2 is a "bad" node, since it forwards only 20% of the packets that should be forwarded. Node 5 has never interacted with node 3 and is ignorant about its behavior. The simulation show that the most appropriate values for the trustworthiness parameters are $x = \sqrt{2}$ and $y = \sqrt{9}$. These parameter values will be used to map trust and confidence to trustworthiness values in our later simulations.

Note that node 5 correctly assigns a trustworthiness value of 0.90 to node 4 and an opinion value of 0.20 to node 2 even after a small number of windows. A trustworthiness value of 0.38 is assigned to node 3 and to all other nodes that node 5 is ignorant about. When the trustworthiness parameters are chosen as $x = y = \sqrt{2}$ (i.e., the ellipse becomes a circle), node 5 places an unreasonably high opinion value on node 2 and an unreasonably low trustworthiness on node 3. Note that when the trustworthiness parameters are set to $x = \sqrt{2}$ and $y = \sqrt{12}$, node 5 penalizes nodes 4 and 2 more than it should.

Calculation of Opinion

In order to demonstrate Hermes's ability to adapt to changes in the node behaviors, we simulate the network topology of Fig. 3.11 with the same flow as before, i.e., node 5 sends 20 packets per window for 30 windows. However, in the present scenario, the intermediate nodes 4 and 1 forward 90% of the packets sent by node 5 in each window, node 8 forwards 90% of the first 100 packets sent to it and 20% of the remaining packets sent to it. Finally,

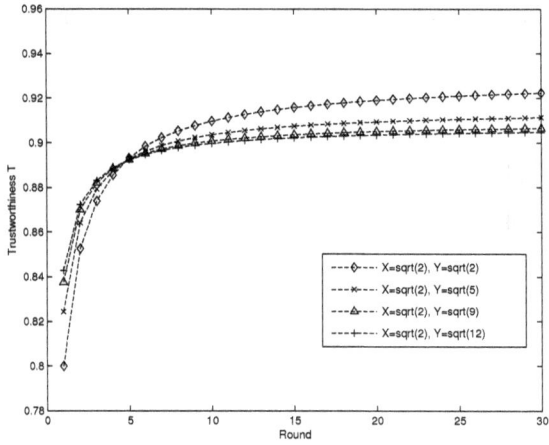

Figure 3.13: Opinion/trustworthiness value $P_{5,4} = T_{5,4}$ for different trustworthiness parameter values.

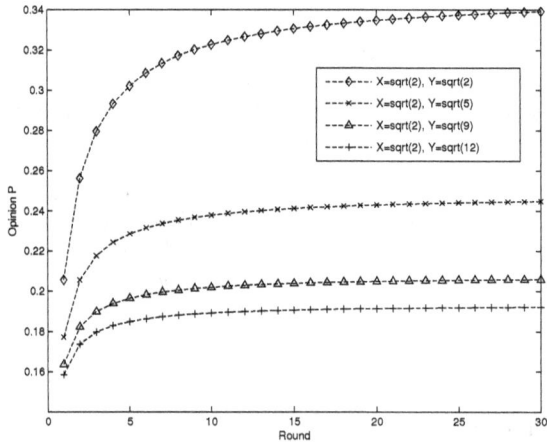

Figure 3.14: Opinion value $P_{5,2}$ for different trustworthiness parameter values.

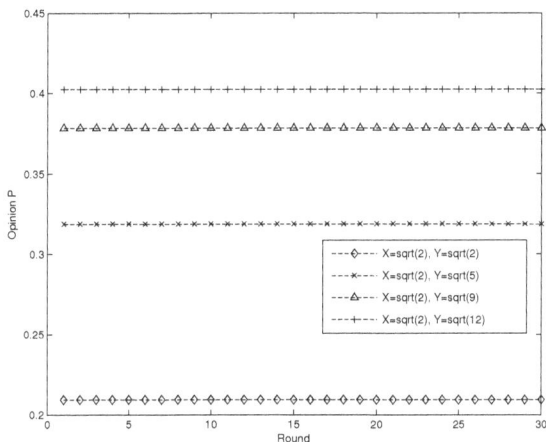

Figure 3.15: Opinion value $P_{5,3}$ for trustworthiness parameter values.

intermediate node 2 exhibits malicious behavior by forwarding only 20% of the packets it receives. The trustworthiness parameters are set as follows: $x = \sqrt{2}$ and $y = \sqrt{9}$.

The opinions that node 5 places on the intermediate nodes over 30 windows when the window size is 20, 50, and 100, respectively, are shown in Figs. 3.16, 3.17, and 3.18, respectively. From Fig. 3.16, we can make the following observations:

1. Node 5 correctly computes an opinion for node 4 of value $P_{5,4} = T_{5,4} = 0.91$. The opinion node 5 has for node 4 is based on the direct observations of its packet forwarding behavior.

2. Node 5 computes an opinion for node 1 of value $P_{5,1} = 0.82 = T_{5,4} \cdot T_{4,1}$.

3. Node 5 detects the change in the behavior of node 8. At the end of window 5, node 5 calculates an opinion for node 8 of value $P_{5,8} = 0.75$. From window 6 onwards, the opinion value $P_{5,8}$ drops to 0.23. The change in the node behavior of node 8 is detected within one window.

4. Up until the fifth window, node 5 considers node 8 "trustworthy" ($P_{5,8} = 0.75$) and accepts its recommendations for node 2. As a result, node 5 correctly assigns an opinion value of $P_{5,2} = 0.22$ to node 2, which always exhibits malicious behavior. From window 6 onwards, node 5 assigns a small opinion value to node 8, and does not accept its recommendations. The opinion value node 5 has for node 2 drops to 0.09, as expected.

5. Node 5 assigns the correct opinion values to the intermediate nodes after a single observation window.

38

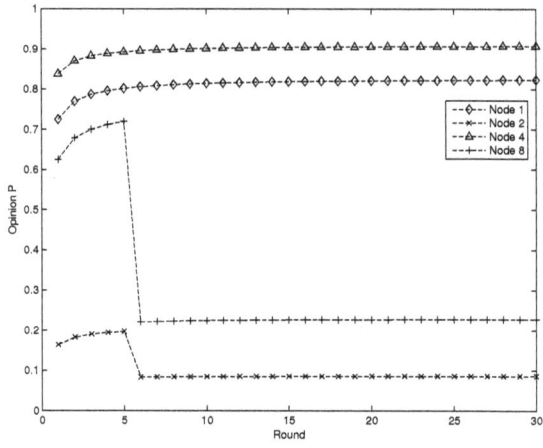

Figure 3.16: Opinion values P that node 5 places on the intermediate nodes when $W = 20$.

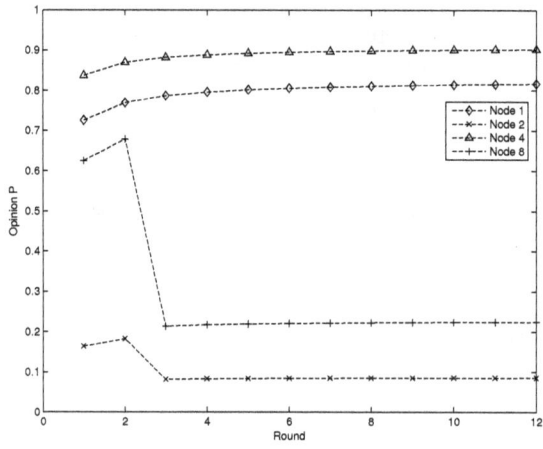

Figure 3.17: Opinion values P that node 5 places on the intermediate nodes when $W = 50$.

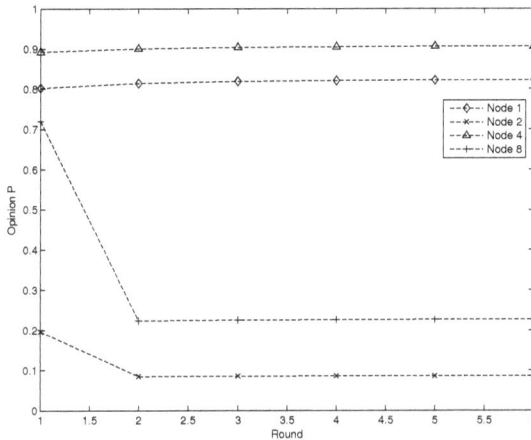

Figure 3.18: Opinion values P that node 5 places on the intermediate nodes when $W = 100$.

Figs. 3.17 and 3.18 are similar to Fig. 3.16 with the difference that the size of the window W is increased to 50 and 100, respectively. Hence, for the same simulation time, the trust, confidence and trustworthiness are computed fewer times, and the number of windows is smaller than in Fig. 3.16 even though the number of packets sent is the same, i.e., 600 packets.

In Fig. 3.16, 20 packets per window are sent over a span of 30 windows. At window 6, node 8 starts misbehaving. In Fig. 3.17, 50 packets per window are sent over a span of 12 windows. Node 8 starts misbehaving at window 2. In Fig. 3.18, 100 packets per window are sent for 6 rounds and node 8 starts misbehaving at window 1. As expected, the smaller the time window W: (1) the sooner a change in a node's behavior is detected and (2) the sooner the source node (in this case, node 5) computes its first opinion about the intermediate nodes. Thus, we see a tradeoff between speed of detection and processing overhead.

Routing Opinion

In the third simulation scenario, five traffic flows are established in the network as follows:

- flow 1 along the path $\{7, 2, 8, 1, 4, 5\}$;

- flow 2 along the path $\{3, 4, 1, 6\}$;

- flow 3 along the path $\{4, 1, 8, 10\}$;

- flow 4 along the path $\{5, 4, 1, 9, 2\}$;

- flow 5 along the path $\{10, 2, 9, 1, 4, 3\}$.

40

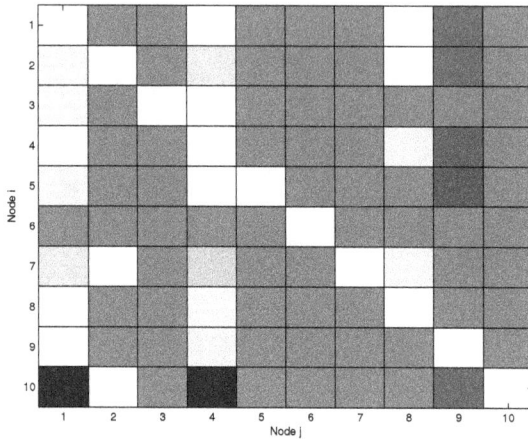

Figure 3.19: Opinion values $P_{i,j}$ in gray-scale.

Node 9 acts maliciously, forwarding only 20% of the packets it should be forwarding. All other nodes forward 90% of the packets they should be forwarding. The source node of each flow sends 20 packets per window over the course of 30 rounds.

Fig. 3.19 illustrates the opinion values, $P_{i,j}$, that node i places on node j with a gray-scale representation. A black color implies an opinion value of 0, whereas white represents an opinion value of 1, while intermediate values are represented by different shades of gray. Fig. 3.20 shows the corresponding numerical opinion values. One can verify that the source and intermediate nodes of these 5 flows have formed the correct opinion about the other nodes. Recall that node 9 is malicious, and is part of flows 4 and 5. Nodes upstream from node 9 in these two flows nodes, i.e., nodes 5, 4, 1, 10, and 2, have formed the correct opinion for it. The corresponding cells of the ninth column of Fig. 3.19 are the darker. The cells of value 0.3784 correspond to links between nodes that have never interacted.

We now investigate three different routing scenarios described as follows:

1. Node 2 does not initially start a session, but has been an intermediate node for one of the five previous flows. Then node 2 requests a route to node 1. The implemented protocol finds two possible routes: route $R_1 = \{2, 9, 1\}$ and $R_2 = \{2, 8, 1\}$. From Fig. 3.20, we can see that node 2 has formed opinions for nodes 9 and 8 already. Node 2 calculates, using equation (3.23), the routing opinion values $V_{R_1} = (P_{2,9})^{1/1} = 0.28$ and $V_{R_2} = (T_{2,8})^{1/1} = 0.91$. The route with the highest routing opinion is chosen to route the packets to the destination node 1. Thus, node 2 successfully avoids the route that includes the malicious node 9.

2. Node 4 has established a session already. Then, node 4 requests a route to node 2. The routing protocol finds two possible routes: $R_1 = \{4, 1, 8, 2\}$ and $R_2 = \{4, 1, 9, 2\}$.

41

```
                                    Node j
 |_____1_____2_____3_____4_____5_____6_____7_____8_____9_____10___|
 |  1 | 1    | 0.38 | 0.38 | 0.91 | 0.38 | 0.38 | 0.38 | 0.91 | 0.28 | 0.38 |
 |  2 | 0.82 | 1    | 0.38 | 0.75 | 0.38 | 0.38 | 0.38 | 0.91 | 0.28 | 0.38 |
N|  3 | 0.82 | 0.38 | 1    | 0.91 | 0.38 | 0.38 | 0.38 | 0.38 | 0.38 | 0.38 |
o|  4 | 0.91 | 0.38 | 0.38 | 1    | 0.38 | 0.38 | 0.38 | 0.82 | 0.25 | 0.38 |
d|  5 | 0.82 | 0.38 | 0.38 | 0.91 | 1    | 0.38 | 0.38 | 0.38 | 0.23 | 0.38 |
e|  6 | 0.38 | 0.38 | 0.38 | 0.38 | 0.38 | 1    | 0.38 | 0.38 | 0.38 | 0.38 |
 |  7 | 0.75 | 0.91 | 0.38 | 0.68 | 0.38 | 0.38 | 1    | 0.82 | 0.38 | 0.38 |
i|  8 | 0.91 | 0.38 | 0.38 | 0.82 | 0.38 | 0.38 | 0.38 | 1    | 0.38 | 0.38 |
 |  9 | 0.90 | 0.38 | 0.38 | 0.81 | 0.38 | 0.38 | 0.38 | 0.38 | 1    | 0.38 |
 | 10 | 0.09 | 0.91 | 0.38 | 0.09 | 0.38 | 0.38 | 0.38 | 0.38 | 0.25 | 1    |
 |_____|
```

Figure 3.20: Opinion values $P_{i,j}$ in numerical values.

From Fig. 3.20, we can see the opinions that node 4 has formed for nodes 1, 8, and 9. Node 2 calculates the following routing opinion values, using equation (3.23): $V_{R_1} = (P_{4,1} \cdot P_{4,8})^{1/2} = (0.91 \cdot 0.82)^{1/2} = 0.86$, $V_{R_2} = (P_{4,1} \cdot P_{4,9})^{1/2} = (0.91 \cdot 0.25)^{1/2} = 0.47$. Thus, route R_1 is selected to route packets from node 4 to node 2. This choice of routes successfully avoids the route that contains the malicious node 9.

3. Node 10 requests a route to node 9. The routing protocol finds two possible routes: $R_1 = \{10, 8, 9\}$ and $R_2 = \{10, 2, 9\}$. From Fig. 3.20, node 2 calculates the following routing opinion values, using equation (3.23): $V_{R_1} = (P_{10,8})^{1/1} = 0.38$ and $V_{R_2} = (P_{10,2})^{1/1} = 0.91$. In this case, route R_2 is selected.

In the above scenario, the Hermes scheme is able to determine the opinion metrics with sufficient accuracy to enable a choice of the best routes with respect to reliable packet delivery. We point out that the accuracy of the opinion metrics should improve further in the presence of a larger set of active flows.

3.8 Hermes Summary

We proposed Hermes, a quantitative trust establishment framework for MANETs, which is designed to improve the reliability of packet forwarding over multi-hop routes in the presence of potentially malicious nodes. Using a Bayesian framework, two metrics are defined: trust and confidence, which are computed based on the empirical first-hand observations of packet forwarding behavior by neighbor nodes. Trust characterizes the belief in the reliability of a neighbor node with respect to packet forwarding. The confidence value associated with a given trust value characterizes the statistical reliability of the trust value. Trust and confidence are mapped into a "trustworthiness" metric, which captures the impact of trust and confidence in a single value.

The concept of trustworthiness, defined only between neighbor nodes, is then extended to the notion of an *opinion* that a given node has for any arbitrary node. The opinion metric can be incorporated into MANET routing protocols such as DSR or AODV to improve the reliability of packet delivery in a transparent manner. A windowing scheme is proposed to expire old observation data in order to improve the accuracy of the opinion metric. The overhead imposed by the Hermes scheme is mainly computational. Nodes following the Hermes scheme collect statistics based on first-hand observations of packet transmissons on the wireless broadcast channel and compute the trust metrics. The communication overhead due to the propagation of second-hand trust information can be minimized by piggybacking trust information onto the routing control packets.

We remark that the Hermes scheme may stimulate selfish nodes not to forward packets in order to conserve battery power. A scheme for punishing such nodes is presented in Section 5.3. The objective of the basic Hermes framework is to distinguish the subset of nodes that can reliably be characterized as being "good" from among the set of all nodes.

A probabilistic attacker model was proposed to characterize the security properties of Hermes. Our simulation experiments demonstrate the effectiveness of the Hermes framework in distinguishing among bad and good nodes as well as in the selection of more "trustworthy" routes for reliable packet delivery. In Chapter 4, we investigate extensions to the Hermes framework to deal with the behavior of nodes that propagate invalid trustworthiness information.

Chapter 4: Robust Trust Propagation

The focus of this Chapter is to extend the Hermes scheme to address an attacker model where nodes can exhibit malicious behaviors independently, i.e., failure to forward packets is independent of the correctness with which trustworthiness values are propagated about other nodes. Another extension of Hermes discussed in this Chapter is a mechanism for deriving trustworthiness values for non-neighbor nodes based on first-hand information from acknowledgements, as opposed to relying on second-hand recommendations alone. Most of the material in this Chapter also appears in [76, 77].

4.1 Overview

In this Chapter, we present a robust, cooperative trust establishment scheme, called *E-Hermes* (Extended-Hermes), which enables a given node to identify other nodes in terms of how "trustworthy" they are with respect to reliable packet delivery. The proposed scheme is cooperative in that nodes exchange information in the process of computing trust metrics with respect to other nodes. At the same time, the scheme is robust in the presence of malicious nodes that propagate false trust information.

The proposed scheme extends Hermes, which was presented in Chapter 3. In Hermes, trust establishment of non-neighbor nodes relies on the second-hand trust information obtained from the propagation of recommendations. A drawback of the Hermes scheme is that it lacks robustness with respect to the propagation of trust information among nodes. In particular, the scheme is vulnerable to attacks by nodes that propagate erroneous trust information in the network. The trust establishment scheme proposed in the present Chapter avoids such attacks by extending the notion of first-hand evidence among neighbor nodes to non-neighbor nodes by employing a protocol involving acknowledgements. Thus, a given node need not rely on second-hand trust information to compute trust metrics with respect to a non-neighbor node. Nevertheless, the sharing of second-hand trust information accelerates the convergence of trust computations.

We note that for a trust establishment scheme to be effective, it must be capable of adapting to the dynamic changes in the topology of a MANET. This issue is addressed in the Hermes scheme via a windowing mechanism, presented in Section 3.4.4, that systematically expires old observation data to maintain the accuracy of computed trust metrics in a dynamic network environment.

The main contribution of the present Chapter is a robust cooperative trust establishment scheme for MANETs, which addresses the following threats:

- packet forwarding misbehavior,

- propagation of false trust information.

The proposed E-Hermes scheme obtains first-hand trust information with respect to non-neighbor nodes and combines this information with second-hand trust information to accelerate the establishment of trust in an ad hoc network. The key novel components of the proposed trust establishment scheme are an acknowledgement scheme for first-hand trust information with respect to non-neighbor nodes and a recommendation scheme that is robust against the propagation of false trust information by malicious nodes. The proposed scheme has the following features: (1) nodes form more accurate opinions for other network node over time as statistical evidence is accumulated; (2) separate opinions are maintained with respect to packet to packet forwarding and trust propagation behaviors; and (3) the scheme is resilient to both individual and colluding attacks on packet delivery. Simulation results are presented to demonstrate the effectiveness of the E-Hermes scheme in distinguishing between malicious vs. non-malicious nodes in a variety of scenarios involving nodes that are malicious both with respect to packet forwarding and trust propagation.

The remainder of the Chapter is organized as follows. Section 4.2 reviews some background material on trust metrics. Sections 4.3 and 4.4 discuss the core concepts and advances of the Chapter. Section 4.3 discusses a protocol for accumulating trust information and computing trust metrics for non-neighbor nodes via acknowledgements. Section 4.4 describes a scheme for cooperatively sharing trust information among nodes via recommendations. Second-hand trust information is combined with first-hand trust information to derive an opinion metric, which summarizes the trust that a given node attributes for another node. Section 4.5 proposes an authentication scheme for both data packets and control packets used for trust establishment. Section 4.6 discusses the security properties of the trust establishment scheme. Section 4.7 presents results from simulation experiments that demonstrate the robustness and key properties of the proposed trust establishment scheme. Finally, the Chapter is concluded in Section 4.8.

4.2 Trust Metrics

We briefly review the notions of trust, confidence, and trustworthiness introduced in the original Hermes scheme introduced in Chapter 3. Consider a given node that is observed over time with respect to its packet forwarding behavior. Let A denote the cumulative number of packets forwarded correctly and let M denote the cumulative number of packets sent for forwarding by the node up to the current time. Then the trust value, t, assigned to a node is defined as follows:

$$t \triangleq \frac{A}{M}, \tag{4.1}$$

where $0 \leq t \leq 1$. A value of t equal to one indicates absolute trust, whereas a value close to zero indicates low trust. This definition of trust is based on Bayesian statistics [7]. The confidence value, c, associated with the trust value t is defined as follows:

$$c = 1 - \sqrt{\frac{12A(M-A)}{M^2(M+1)}}, \tag{4.2}$$

where $0 \leqslant c \leqslant 1$. A value of c close to one indicates high confidence in the accuracy of the computed trust value t, whereas a value close to zero indicates low confidence. At a given time instant a node can be characterized by a pair (t, c). In particular, node i characterizes its trust in node j by the pair $(t_{i,j}, c_{i,j})$.

The *trustworthiness* metric characterizes a pair (t,c) of trust and confidence values into a single value to facilitate trust-based decisions. The trustworthiness associated with a pair (t,c) is defined as (see also (3.7))

$$T(t,c) \triangleq 1 - \frac{\sqrt{(t-1)^2 + r^2(c-1)^2}}{\sqrt{1+r^2}}, \qquad (4.3)$$

where r is a parameter that determines the relative importance of the trust value t vs. the confidence value c. The "default" value of trustworthiness is defined as

$$T_{def} \triangleq T(0.5, 0), \qquad (4.4)$$

which represents the trustworthiness value assigned to a node when its assigned trust and confidence values are $t = 0.5$ and $c = 0$, respectively. Thus, the value T_{def} represents ignorance about the trustworthiness of a node. The value T_{def} can be interpreted as an initial threshold for trustworthiness. If the trustworthiness of a node exceeds T_{def}, then the node is considered "trustworthy" or "good". Otherwise, the node is viewed as "untrustworthy" or "bad".

In addition to T_{def} we also define c_{acc} as an *acceptability* threshold with respect to the confidence level. The concept of acceptability is used in calculating second-hand trust information (see Section 4.4.1). The pair (t,c) is *acceptable* if a sufficient amount of observation data has been accumulated such that $c > c_{acc}$. This condition can also be formulated in terms of trustworthiness values by defining the *acceptable trustworthiness* value for trust value t as:

$$T_{acc} = T(t, c_{acc}). \qquad (4.5)$$

Then, the trustworthiness value $T = T(t,c)$ is accepted if $T > T_{acc}$. We remark that each node may choose a different value of c_{acc} to implement its own policy in determining the acceptability of trustworthiness values.

4.3 First-hand Trust Evaluation

In this Section, we present a new scheme for gathering first-hand trust information from non-neighbor nodes. This is an extension over Hermes, which gathers first-hand information only from neighbor nodes. The scheme requires authentication mechanisms for both data and control packets.

4.3.1 Neighbor Nodes

We first review the Hermes approach for establishing trust for neighbor nodes introduced in Section 3.4. In the Hermes scheme, nodes evaluate the trustworthiness of their neighbors by snooping the wireless channel. It is assumed that the nodes are equipped with omnidirectional antennas and that they do not employ dynamic power control. We use the term *fault* to denote an event in which a node fails to forward a packet correctly to its next hop. A fault may occur due to malicious or non-malicious misbehavior of a node. Non-malicious packet forwarding misbehavior may be due to such phenomena as network congestion, node mobility, or node malfunction.

Consider a very simple route $\{x, y, z\}$. In this scheme, a given node x in the network maintains counters M_y and A_y for a neighbor node such as y. We refer to the sets of counters

46

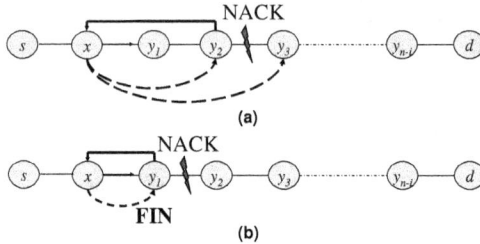

Figure 4.1: Processing of NACKs.

$\{M_y\}$ and $\{A_y\}$ as M-counters and A-counters, respectively. The counter M_y records the total number of packets sent from node x to node y for forwarding to z over an observation window. The counter A_y records the total number of packets forwarded *correctly* (not dropped or misrouted) from node y to node z.

The counters M_y and A_y are updated as follows. Whenever a packet p is forwarded from node x to node y, M_y is incremented by one and a timer is initiated. The timeout interval is set to a value greater than the maximum round-trip time (RTT) between two neighbor nodes in the network. If node x observes a copy of packet p forwarded from node y correctly to the next hop (say node z) before the timer expires the counter A_y is incremented by one. Otherwise, the counter A_y is not updated.

Definition 1. *The expiry of the timer indicates that an incorrect packet forwarding event has occurred. We refer to this event as a* **fault**. *When a fault attributed to node y occurs, the counter M_y is incremented by one and A_y is not updated. In this case, we say that node x* **penalizes** *node y.*

4.3.2 Non-neighbor Nodes

We now generalize the evaluation of first-hand trust to non-neighbor nodes for the case when the underlying routing protocol is based on source routing, such as DSR [72]. In source routing, the complete path taken by the packet is known at the source. To incorporate trust into source routing, nodes must establish trust for non-neighbor nodes. We remark that for distance vector routing protocols, such as AODV [70], it is sufficient to establish trust only for neighbor nodes.

To obtain first-hand information from non-neighbor nodes, we propose an acknowledgement scheme. Consider the topology given in Fig. 4.1. When node x forwards packet p to node y_1, it initiates an *acknowledgement timer* with timeout interval t^{ack} and updates the

47

M-counters for the downstream intermediate nodes as follows:

$$M_{y_i} \leftarrow M_{y_i} + 1, \quad 1 \leq i \leq n - 1. \tag{4.6}$$

The value of timeout interval t^{ack} should be larger than the maximum round-trip propagation time along the given path in the network.

In the case of an acknowledgement (ACK) packet, node x forwards the ACK to its upstream neighbor (either another intermediate node or the source node itself), and updates the A-counters for all of the downstream intermediate nodes as follows:

$$A_{y_i} \leftarrow A_{y_i} + 1, \quad 2 \leq i \leq n - 1, \tag{4.7}$$

which indicates that all of the downstream nodes had correctly forwarded the packet p. Since node y_1 is a direct neighbor of node x, the counter A_{y_1} is updated based on observation of the packet forwarding behavior at the MAC layer as discussed earlier in Section 4.3.1.

Now let us consider the case when an ACK is not received. There are three subcases. First, if an intermediate node along the path R_p fails to receive an ACK packet within the timeout period, it creates a negative acknowledgement (NACK) packet and sends the NACK to its upstream neighbor on the path. The other two subcases are illustrated in Fig. 4.1, in which the dashed arrow from node x to node y indicates that x penalizes y for the fault and the solid arrow indicates the transmission of a NACK. Fig. 4.1(a) illustrates the subcase where node x receives a negative ACK (NACK) from node y_2 and Fig. 4.1(b) illustrates the subcase where node x receives a NACK from node y_1. In the second subcase, y_1 is what we call a first intermediate node (FIN). The FIN node is the first node that is downstream from the recipient of the NACK.

We discuss the two subcases as follows:

1. *NACK originating from node y_i, $2 \leq i \leq n - 1$*: In this case, node x infers that a fault occurred on the link (y_i, y_{i+1}), but cannot identify which of the two nodes y_i and y_{i+1} caused the fault (y_2 or y_3 in Fig. 4.1(a)). Therefore, node x penalizes both nodes for the fault. To avoid unnecessarily penalizing the nodes downstream from y_{i+1}, the M-counters for these nodes are decremented by one as follows:

$$M_{y_j} \leftarrow M_{y_j} - 1, \quad i + 2 \leq j \leq n - 1. \tag{4.8}$$

On the other hand, the intermediate nodes $y_1, \cdots, y_{i-1}$ should receive credit for correctly forwarding the packet. This is done by incrementing the corresponding A-counters by one:

$$A_{y_j} \leftarrow A_{y_j} + 1, \quad 1 \leq j \leq i - 1. \tag{4.9}$$

2. *NACK originating from FIN node y_1*: If node x had previously observed at the MAC layer that node y_1 correctly forwarded packet p, then node x assumes that node y_2 failed to forward the packet correctly. In other words, the FIN property of y_1 implies that x can monitor the forwarding behavior of y_1 at the MAC layer and this allows it to isolate the fault to y_2. To avoid penalizing the nodes downstream from y_2, the M-counters for the nodes $y_3, \cdots, y_{n-1}$ are decremented by one:

$$M_{y_i} \leftarrow M_{y_i} - 1, \quad 3 \leq i \leq n - 1. \tag{4.10}$$

On the other hand, if node x had observed at the MAC layer that node y_1 incorrectly forwarded the packet p (see Fig. 4.1(b)), then the nodes downstream from y_1 should not be penalized. Therefore, node x decrements the M-counters for these nodes by one:

$$M_{y_i} \leftarrow M_{y_i} - 1, \quad 2 \leq i \leq n - 1. \tag{4.11}$$

Note that the counters are maintained only for downstream nodes. The reason is that an intermediate node knows the number of packets it receives for forwarding from its upstream node, but is unaware of the number of packets that its upstream node had received for forwarding. We also remark that upon receipt of a NACK, both ends of the associated link are penalized even if only one of the nodes may be responsible for the fault. However, this effect diminishes as more observation data involving the two nodes with respect to different flows is accumulated over time.

4.3.3 Computing Trustworthiness

Given the counters M_y and A_y, maintained for both neighbor and non-neighbor nodes with which a source node interacts, the number of packets forwarded *incorrectly* by node y is given by $B_y \triangleq M_y - A_y$. Then the trust and confidence that x attributes to y over an observation window are given by (cf. (4.1) and (4.2))

$$t_y = t(A_y, B_y) \text{ and } c_y = c(A_y, B_y),$$

from which the trustworthiness value T_y can be computed via (4.3). In the next Section we discuss how trustworthiness calculations from neighbor and non-neighbor nodes are combined to formulate *opinions*.

4.4 Formulation of Opinions for E-Hermes

Node i may need to make routing or other network-related decisions that involve nodes, for example, a node m for which trustworthiness value $T_{i,m}$ is below T_{acc}. In this case, second-hand trustworthiness values from third-party nodes are incorporated to form an opinion about node m. The propagation of trustworthiness information to form opinions is accomplished through *recommendations*.

4.4.1 Processing Recommendations

Definition 2. *A* **recommendation** *by node j on node m is an assertion by j of the trustworthiness, which it has for node m (denoted as $T_{j,m}$). Node j is thus the* **recommender**.

Node i seeks recommendations on a node m when the trustworthiness value it has computed for m is below T_{acc} (see Section 4.2). Node i discriminates among multiple recommenders by evaluating a metric called *recommender trustworthiness*.

Definition 3. *Recommender trustworthiness $T_{i,j}^R$ is the trustworthiness that node i places on recommender node j as a measure of how reliably node j propagates trustworthiness information.*

Definition 4. *A node j is* **considered a good recommender** *by node i when the recommender trustworthiness $T_{i,j}^R$ that i places on recommender j exceeds T_{def}.*

Definition 5. *A node j is **considered a bad recommender** by node i when the recommender trustworthiness $T_{i,j}^R$ that i places on recommender j is smaller than T_{def}.*

Consider a scenario where node i asks a set of nodes D for their recommendations for node m. Recommendations are sought when a node wishes to establish a route in which some of the nodes have a trustworthiness value smaller than T_{acc}. The *recommender set* D is chosen from among all nodes in the network in the following order of priority: (i) good recommenders, (ii) nodes for which the recommender trustworthiness $T^R < T_{acc}$, and (iii) all other bad recommenders. We remark that bad recommenders may be chosen as part of the recommender set in order to update their recommender trustworthiness values. The recommender set D is limited to a size d to limit the communication overhead. No mechanisms are in place to obligate nodes to respond to recommendation requests. We assume that node i will receive $f \leq d$ recommendations due to network conditions or lack of willingness to respond to the request. Additionally, when node j places on node m trustworthiness smaller than T_{acc}, j does not reply to node i's recommendation request. Recommendations are authenticated with a message authentication code (MAC) computed using the shared keys between the source s and the destination d of the request or the reply.

After receiving a set $R_m = \{T_{j,m} : j \in D\}$ of recommendations for node m, node i performs the following steps. If the trustworthiness value $T_{i,m}$ is smaller than T_{acc}, node i calculates a "temporary" trustworthiness value $\tilde{T}_{i,m}$, which is taken as the maximum trustworthiness value $T_{j,m}$ among the recommenders $j \in D$, i.e.,

$$\tilde{T}_{i,m} = \max\{T_{j,m} : j \in D\}. \tag{4.12}$$

The value $\tilde{T}_{i,m}$ is used for routing or any other network-related decisions until subsequent updates lead the value of $T_{i,m}$ exceeding T_{acc}.

When $T_{i,m} > T_{acc}$, the trustworthiness of the recommenders $j \in D$ can be evaluated. This is done by performing the following *recommender's test* or *RC-test*:

$$\text{RC-test}: |T_{i,m} - T_{j,m}| \leq \eta,$$

where $\eta \in (0, 1)$ is a threshold value. The RC-test succeeds when the recommended trustworthiness value is close to the first-hand trustworthiness value as defined by the threshold η. Otherwise, the test fails. The outcome of each RC-test for recommender j is used to update counters A^R and M^R, where A^R counts the number of times for which the RC-test succeeds and M^R counts the total number of times that the RC-test is applied. The A^R and M^R counters are then used to calculate the *recommender trustworthiness* $T_{i,j}^R$ according to the trustworthiness formulas (4.1)- (4.3).

A node j declines to submit a recommendation for node m to node i when m is the FIN node of j and $\eta \cdot 100\%$ of the control packets sent from m to j for a given flow are NACKs. As discussed in Section 4.3.2, when node m sends a NACK upstream, the source node i attributes the fault both to m and its downstream neighbor. On the other hand, since j is a neighbor of m, it can isolate the fault either to node m or its downstream neighbor. In this case, the trustworthiness that i calculates for m, $T_{i,m}$, and the trustworthiness that j calculates for m, $T_{j,m}$, could be significantly different when m is actually a good node. Thus, the RC-test would fail for node j even though it may in fact be a good recommender.

Algorithm 1 summarizes the procedure that node i executes to process recommendations for a node m. The calculation of opinion $P_{i,m}$ is discussed in Section 4.4.2.

Algorithm 1 Processing of recommendations for node m

    choose recommender set D
    obtain recommendations for nodes in D
    if $T_{i,m} < T_{accept}$ **then**
        $T_{i,m}^{tmp} \leftarrow \max\{T_{j,m} : j \in D\}$
    end if
    run RC-test for recommendations $T_{j,m}$, $\forall j \in D$
    update recommender trustworthiness $T_{i,j}^{R}$, $\forall j \in D$
    calculate opinion $P_{i,m}$

4.4.2 Calculation of Opinion

We generalize the notion of trustworthiness to the concept of *opinion*, which incorporates second-hand trustworthiness values from third-party nodes. We denote the opinion that node i has for node m by $P_{i,m}$. The definition for the opinion that any node i has for another node m is given as follows:

$$P_{i,m} \triangleq \max_{j \in \Gamma}\{\omega_{i,j} T_{j,m}\}, \text{ for } P_{j,m} \neq T_{def}, \qquad (4.13)$$

where

$$\omega_{i,j} = \begin{cases} T_{i,j}^{R}, & i \neq j, \\ 1, & i = j. \end{cases} \qquad (4.14)$$

and Γ is the set of recommenders in D that have passed the RC-test.

Nodes are judged to be *good* or *bad* on the basis of the opinion value.

Definition 6. *A node j is **considered good** by node i when the opinion $P_{i,j} > T_{def}$.*

Definition 7. *A node j is **considered bad** by node i when the opinion $P_{i,j} < T_{def}$.*

4.5 Authentication of Packets

Authentication of every data, recommendation, ACK, and NACK packet is required to protect the network against modification and impersonation attacks. The E-Hermes scheme requires all nodes to verify the authenticity of ACK/NACK packets received from downstream nodes in order to draw reliable conclusions about their packet forwarding behaviors. Hash chains provide a convenient mechanism for hop-by-hop authentication verification of ACK/NACK packets to be performed by intermediate nodes.

We assume that the nodes have already established a set of pairwise keys using a key management protocol [9,49,64,65]. If a secure routing protocol is in place, the keys established for secure routing can be used to secure the E-Hermes scheme. Let $K_{i,j}$ denote the shared key between node i and node j. Consider a path $R = \{s, a_1, a_2, \cdots, a_{n-1}, a_n = d\}$, where $n \geq 2$, from source node s to destination node d. Let k denote the sequence number of a given data packet that is forwarded along the path R.

4.5.1 Data and Recommendation Packets

As in [61], the authentication field, $\mathcal{A}$, of a data packet of data field $\mathcal{D}$ sent along route R, consists of a sequence of message authentication codes (MACs):

$$\mathcal{A} = [\mathcal{M}_n, \mathcal{M}_{n-1}, \cdots, \mathcal{M}_1].$$

The MACs are defined as follows:

$$\mathcal{M}_n = f(K_{s,a_n}, \mathcal{D}),$$

and for $i = 1, \cdots, n - 1$:

$$\mathcal{M}_i = f(K_{s,a_i}, [\mathcal{D}, \mathcal{M}_n, \cdots, \mathcal{M}_{i+1}]),$$

where $f(K, \mathcal{X})$ denotes the function that produces a MAC from the key K and data $\mathcal{X}$. The authentication field allows each intermediate node to authenticate the packet and protects against malicious intermediate nodes that try to tamper with the MAC field of a downstream node. In the E-Hermes scheme, the intermediate nodes along the route need to be able to authenticate data packets in order to collect packet statistics to derive first-hand trust information.

Recommendation request and reply packets are not used to collect first-hand trust information. Therefore, for recommendation packets, it suffices for the authentication field to consist only of a single MAC computed using the shared key between the recommender and the source s of the recommendation request.

4.5.2 Control Packets

In this Section, we propose an extension of the hash chain mechanism proposed in [78] to provide authentication of ACK and NACK packets in the E-Hermes scheme. The authentication fields of each ACK/NACK packet are designed to satisfy four properties: (i) they are computationally impractical to forge, (ii) if an ACK/NACK verifies at one non-faulty node on the path, it also verifies at all non-faulty routers on the path, (iii) they authenticate the identity of the nodes that appended them, and (iv) they authenticate the packet content (i.e., whether the packet is an ACK or an NACK). The first three properties are also satisfied by the scheme proposed in [78], the fourth property is the result of our proposed extension as discussed below.

A one-way hash function $h(\cdot)$ and hash chains of length three, associated with each data packet, are used to guarantee these properties. We propose that for packet k and intermediate node a_i, a hash chain is used to authenticate ACK packets traveling upstream on route R. Let $\alpha_i^0(k)$ denote the initial element of the "ACK" hash chain for node a_i for $i = 1, \cdots, n$. The hash chain element $\alpha_i^0(k)$ is constructed by concatenating the key $K_s^{a_i}$, the sequence number k, and the element 0. The second and third elements in the ACK hash chain associated with packet k and node a_i are

$$\alpha_i^1(k) \triangleq h[\alpha_i^0(k)] \text{ and } \alpha_i^2(k) \triangleq h[\alpha_i^1(k)],$$

respectively.

In a similar way, a three-element hash chain associated with packet k and node a_i is defined for the authentication of NACK packets for $i = 1, \cdots, n - 1$. Note that a "NACK" hash chain element is not required for the destination node a_n, since it never transmits NACK packets. Let $\eta_i^0(k)$ denote the initial element of the NACK hash chain for node a_i. The hash chain element $\eta_i^0(k)$ is constructed by concatenating the key $K_s^{a_i}$, the sequence number k, and the element 1. The second and third elements in the ACK hash chain associated with packet k and node a_i are

$$\eta_i^1(k) \triangleq h[\eta_i^0(k)] \text{ and } \eta_i^2(k) \triangleq h[\eta_i^1(k)],$$

respectively.

When node s transmits data packet k along route R, it concatenates the third elements of the ACK and NACK hash chains associated with the intermediate nodes, i.e.,

$$\alpha_1^2(k), \alpha_2^2(k), \cdots, \alpha_n^2(k),$$
$$\eta_1^2(k), \eta_2^2(k), \cdots, \eta_{n-1}^2(k).$$

Thus, $2n - 1$ hash chain elements are concatenated to each data packet as part of the payload $\mathcal{D}$. As packet k is forwarded along the path R, each intermediate node a_i $(1 \leq i \leq n - 1)$ extracts and stores the $2n - 1$ hash chain elements (actually, only the hash chain elements corresponding to the downstream nodes are required).

Packet k is protected using hop-by-hop authentication with MACs, as discussed in Section 4.5.1. Packet k is considered to be delivered successfully if the source s receives and verifies the authentication field of the ACK packet corresponding to packet k from its downstream neighbor node a_1 within a predetermined timeout period. This ACK packet is created initially by the destination node d by concatenating the element $\alpha_n^1(k)$ to form the payload as follows:

$$[\text{ACK} | \alpha_n^1(k)]$$

Node d can compute the element $\alpha_n^1(k)$, since it knows the shared secret key $K_s^{a_n}$, and the data packet sequence number k. The ACK packet is then forwarded along the reverse path to node a_{n-1}. If the ACK packet reaches node a_{n-1} successfully, node a_{n-1} verifies the authenticity of the ACK packet by applying the hash function $h(\cdot)$ to the concatenated element $\alpha_n^1(k)$ and comparing the result with the previously stored element $\alpha_n^2(k)$. The ACK packet authenticates successfully if and only if a match occurs.

If the authentication is successful, node a_{n-1} concatenates the element $\alpha_{n-1}^1(k)$ to the ACK packet and forwards the packet

$$[\text{ACK} | \alpha_n^1(k), \alpha_{n-1}^1(k)]$$

to the next node on the reverse path, i.e., node a_{n-2}. The same procedure is followed by the remaining intermediate nodes on the reverse path to the source s. Thus, in the case of a successful packet delivery, node s receives the packet

$$[\text{ACK} | \alpha_n^1(k), \alpha_{n-1}^1(k), \cdots, \alpha_1^1(k)]$$

Node s verifies the authenticity of the ACK packet by applying the hash function $h(\cdot)$ to the concatenated elements $\alpha_i^1(k)$, $(1 \leq i \leq n)$ and comparing the result with the elements $\alpha_i^2(k)$ $(1 \leq i \leq n)$. The ACK packet is successfully authenticated if and only if a match occurs.

If any of the intermediate nodes, say node a_i $(1 \leq i \leq n - 1)$ on the reverse path fails to receive an authenticated ACK packet from its downstream neighbor within a predefined timeout period, it creates a NACK packet, concatenated with the NACK hash chain element $\eta_i^1(k)$ as follows:

$$[\text{NACK}|\eta_i^1(k)]$$

This NACK packet is then sent to node a_{i-1}. Node a_i can compute the element $\eta_i^1(k)$, since it knows the shared secret key $K_s^{a_i}$, and the data packet sequence number k. The NACK packet is then forwarded along the reverse path to node a_{n-1}. Similarly with the case of an ACK receipt, if the NACK packet reaches node a_{i-1} successfully, node a_{i-1} verifies the authenticity of the NACK packet by applying the hash function $h(\cdot)$ to the concatenated element $\eta_i^1(k)$ and comparing the result with the previously stored element $\eta_i^2(k)$. The NACK packet is authenticated successfully if and only if a match occurs. If the authentication is successful, node a_{i-1} concatenates the element $\eta_{i-1}^1(k)$ to the NACK packet and forwards the packet

$$[\text{NACK}|\eta_i^1(k), \eta_{i-1}^1(k)]$$

to the next node on the reverse path, i.e., node a_{n-2}. The same procedure is followed by the remaining intermediate nodes on the reverse path to the source s. Thus, in the case of a successful packet delivery, node s receives the packet

$$[\text{NACK}|\eta_i^1(k), \eta_{i-1}^1(k), \cdots, \eta_1^1(k)]$$

and checks its authenticity.

The security of the scheme relies on (1) the secrecy of the keys and (2) the one-way property of hash function $h(\cdot)$, which defines that given content y and a hash function $h(\cdot)$ it is impossible to derive any content x such that $h(x) = y$. In particular, node a_i cannot derive the elements $\alpha_j^0(k)$, $\alpha_j^1(k)$ $(1 \leq j \leq n, j \neq i)$ and $\eta_j^0(k)$, $\eta_j^1(k)$ $(1 \leq j \leq n - 1, j \neq i)$ for any other node a_j, (1) because of the secrecy of the symmetric secret key $K_s^{a_j}$ and (2) since it is impossible to derive any x such that $h(x) = \alpha_j^2(k)$ or $h(x) = \eta_j^2(k)$.

Upon the successful receipt of ACK or NACK packets, node a_i $(1 \leq i \leq n - 2)$ updates M-counters and A-counters, as discussed in Section 4.3.2. Then, these counters are used to calculate the trust, confidence and trustworthiness that node a_i attributes to each of its downstream nodes over an observation window, as described in Section 4.3.3.

Note that in our proposed scheme $\alpha_i^1(k)$ $(1 \leq i \leq n)$ elements are appended to ACK packets for authentication, whereas $\eta_i^1(k)$ $(1 \leq i \leq n - 1)$ elements are appended to NACK packets for authentication. On the contrary, in [78], elements $r_i^1(k)$ $(1 \leq i \leq n)$ are appended to both ACK and NACK packets for authentication. Similarly to our scheme, the initial hash chain elements $r_i^0(k)$ are constructed by concatenating the key $K_s^{a_i}$ and the sequence number k, whereas $r_i^1(k)$ and $r_i^2(k)$ are the second and third elements of the hash chains respectively. Nonetheless, the latter approach does not satisfy the following requirement: the authentication fields of the ACK/NACK packets must authenticate the packet content (i.e., whether the packet is an ACK or an NACK).

More precisely, the scheme of [78] is vulnerable to the following attack. Consider the path $R = \{s, a_1, a_2, a_3, a_4 = d\}$. Suppose that destination d sends an ACK and authenticates it by appending the element $r_4^1(k)$. Similarly, nodes a_3 and a_2 properly append $r_3^1(k)$ and $r_2^1(k)$ respectively to the successfully received ACK packet and forward it to node a_1. Assume a_1 is a malicious node and (i) drops the received ACK, (ii) discards the elements $r_4^1(k)$ and $r_3^1(k)$, but keeps $r_2^1(k)$ from node a_2, (iii) creates a NACK, to which (iv) it appends its own authentication element $r_1^1(k)$ and $r_2^1(k)$ from node a_2:

$$[\text{NACK}|r_2^1(k), r_1^1(k)]$$

The created NACK is forwarded to the source s. Upon the receipt of this NACK, the source will believe that node a_2 constructed a NACK for link (a_2, a_3) and nodes a_2, a_3 will be erroneously penalized (see Section 4.3).

In our scheme, the malicious node a_1 cannot launch the aforementioned attack, because it cannot create a valid NACK packet and attribute it to node a_2. To understand this, let us consider the same scenario, while our authentication scheme is applied. Again, suppose that the destination d sends an ACK and authenticates it by appending the element $\alpha_4^1(k)$. Similarly, nodes a_3 and a_2 properly append $\alpha_3^1(k)$ and $\alpha_2^1(k)$ respectively to the successfully received ACK packet and forward it to node a_1. Assume a_1 is a malicious node and (i) drops the received ACK, (ii) discards the elements $\alpha_4^1(k)$ and $\alpha_3^1(k)$, but keeps $\alpha_2^1(k)$ from node a_2, (iii) creates a NACK, to which (iv) and appends its own authentication element $\eta_1^1(k)$ and $\alpha_2^1(k)$ from node a_2:

$$[\text{NACK}|\alpha_2^1(k), \eta_1^1(k)]$$

The created NACK is forwarded to the source s, which sees the inconsistency of the authenticator elements in the received NACK packet. The element $\alpha_2^1(k|0)$ authenticates an ACK, whereas the element $\eta_1^1(k|1)$ authenticates a NACK. Thus, the source correctly identifies that the inconsistency appears between nodes a_1 and a_2 and penalizes them.

4.6 Security Evaluation of E-Hermes

4.6.1 Attacker Model

We assume an attacker model in which a node may drop, misroute or replay data packets that it is supposed to forward under a given routing protocol. A node that performs this type of attack with a certain statistical regularity is referred to as a *bad node*. A node that forwards the majority of its packets correctly, with statistical regularity, is referred to as a *good node*. Analogously, we define a *bad recommender* as a node that incorrectly propagates recommendations with a certain statistical regularity. Conversely, a node that propagates recommendations correctly, with high statistical regularity, is a *good recommender*.

The above notations can be made more precise by modeling the frequency with which a node causes a fault in terms of probabilities. More specifically, let η_f and η_t denote, respectively, thresholds on the degrees of packet forwarding and trust propagation misbehaviors that can be tolerated in the network. We set $\eta_f = \eta_t = 1 - T_{def}$. Given that for each node $i \in \mathcal{N}$, B_f^i denotes the probability the node i incorrectly forwards a data packet and B_t^i denotes the probability that it incorrectly propagates a recommendation (according to our probabilistic attacker model introduced in Section 3.5.3), we present the following definitions.

Definition 8. *Node i is **good** if $B_f^i < \eta_f$.*

Definition 9. *Node i is **bad** if $B_f^i > \eta_f$.*

Definition 10. *Node i is a **good recommender** if $B_t^i < \eta_t$.*

Definition 11. *Node i is a **bad recommender** if $B_t^i > \eta_t$.*

A useful measure of the performance of the proposed trust establishment scheme is given as follows.

Definition 12. *The **bad node recognition** percentage or BN-recognition is the percentage of the nodes in the network that consider all the bad nodes in the network as bad nodes.*

We shall assume that every node, whether good or bad, forwards ACK or NACK packets corresponding to packets that it has forwarded earlier. This assumption simplifies the security evaluation given below, but does not represent any limitation in the E-Hermes scheme itself. In the E-Hermes framework, a given node X has nothing to gain by failing to forward an ACK or NACK packet associated with a packet that it has forwarded previously. If node X fails to forward a ACK/NACK packet, node X will be penalized by all of the upstream nodes on the associated route as though it had not forwarded the original packet.

4.6.2 Security Properties of E-Hermes

The key security properties provided by the E-Hermes scheme, beyond what is provided in the original Hermes scheme, are summarized as follows:

1. **Ability to capture independence between packet forwarding and trust propagation misbehaviors.** The proposed scheme can handle the dropping and misrouting of packets, as well as the propagation of false opinion values. Even when false trustworthiness values are computed for recommenders, the scheme converges to opinion values that correctly characterize the underlying packet forwarding behavior.

2. **Resilience to the presence of bad nodes and bad recommenders.** Our simulation studies show very few false positives (i.e., a good node is identified as bad) and false negatives (i.e., a bad node is identified as good) even when the proportion of bad recommenders is as high as 90%. Similarly, the scheme performs well even when the proportion of bad nodes is relatively high.

3. **Resilience to attacker placement.** The ACK processing scheme ensures that a bad node X along a route for a given flow is penalized. The upstream neighbor O of node X will also be penalized, since the ACK scheme essentially penalizes the link between the two nodes. However, if the routing of flows is sufficiently diverse, node O will be eventually credited with good behavior. As discussed in Section 3.6.2, randomization can be used to increase the routing diversity. Routing diversity also increases with the degree of mobility in the network. In this sense, the E-Hermes scheme is resilient to the placement of attackers in the network.

The E-Hermes scheme also provides the security properties provided by the original Hermes scheme as listed below. The E-Hermes scheme generally converges faster and is more robust than the original Hermes scheme due to the gathering of first-hand trust information for non-neighbor nodes via the proposed acknowledgement scheme.

1. **Resilience to multiple, concurrent, and colluding attacks.** Three properties in the scheme collectively ensure resilience against multiple, concurrent, and colluding attacks: (i) the ability of a sender to overhear and verify forwarding of its packets by its FIN nodes, (ii) an acknowledgement scheme that conservatively identifies potentially bad nodes, and (iii) a node i evaluates recommendations from another node j on node m always in relation to the first-hand trustworthiness that node i has computed for node m.

2. **Resilience to attack frequency.** Trust information is calculated as the ratio of misforwarded packets to the total number of forwarded packet. The Hermes scheme estimates this ratio via first-hand observation (and the ACK scheme in the case of E-Hermes), regardless of its actual value. We note that the convergence of Hermes scheme with respect to a particular flow may be slowed down by an upstream attacker by reducing the amount of observation data available. However, with more routing diversity, this effect becomes negligible.

3. **Resilience against packet duplication and replay attacks.** Replay and packet duplication attacks are mitigated using sequence numbers at the network layer (see Section 3.4). Such attacks can thus be stopped by the first non-malicious node on the path of a flow. As a result, the nodes further downstream do not receive duplicate packets.

4.6.3 Security Analysis

We analyze the resistance of E-Hermes to 1) incorrect data packet forwarding, and 2) incorrect propagation of trust information attacks. As discussed in Section 4.1, during the data transmission phase authorized or insider nodes may consistently drop, misroute, or replay data packets. The Hermes scheme identifies such misbehaviors in terms of the trustworthiness and opinion metrics, but does not purport to distinguish between malicious or non-malicious misbehaviors. Non-malicious packet forwarding misbehavior may be due to such phenomena as network congestion, node mobility, or node malfunction. Note that we do not distinguish among the various types of data packet forwarding misbehaviors, i.e., packet dropping, misrouting, and replay attacks.

We now consider the response of the E-Hermes schemes in various attack scenarios, with respect to a single flow. As we shall see, in each case, the E-Hermes scheme successfully penalizes the bad nodes and bad recommenders. The upstream neighbor of the bad node will also be penalized even if it happens to be a good node, since the ACK scheme penalizes bad *links* along the route. In general, however, the upstream neighbor will be credited as a good node with respect to other flows. Routing diversity ensures that a good node will be recognized as bad only with low probability.

Bad Nodes

Fig. 4.2 illustrates the response of Hermes to packet forwarding misbehavior from a single bad node, labelled X, on a route $R_1 = \{Y_2, Y_1, o, X, Z_2, \cdots\}$ corresponding to flow f_1. Node X incorrectly forwards data packets on flow f_1 with probability B_f^X, where $0 < B_f^X \leq 1$. Since node o is a neighbor of X, it obtains first-hand information about the packet-forwarding behavior of node X at the MAC layer. The nodes upstream of node o, i.e., nodes Y_1-Y_3, infer

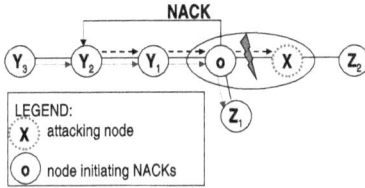

Figure 4.2: Single attacker node.

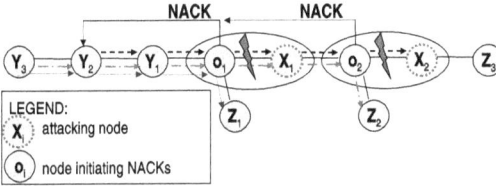

Figure 4.3: Multiple non-neighbor attacker nodes.

first-hand trust information from the NACKs initiated by node o. Since node o is a neighbor of node Y_1, node Y_1 is able to verify the correct forwarding behavior of node o. Thus, upon receiving a NACK from node o, node Y_1 penalizes node X. On the other hand, upon receiving a NACK initiated by node o, nodes Y_1 and Y_2 penalize both nodes o and X. As discused above, node o can be recognized as a good node only through other flows in which node o is not penalized.

Multiple attacker nodes along a route may act independently or collude with each other. In either case, the E-Hermes scheme is able to distinguish good nodes from bad nodes, given a sufficient number of flows along a diverse set of routes. Fig. 4.3 illustrates the situation of packet forwarding misbehavior from multiple non-neighbor attacker nodes X_1 and X_2 on a route $R_1 = \{Y_2, Y_1, o_1, X_1, o_2, X_2, Z_3, \cdots\}$, corresponding to flow f_1. Here, the response of E-Hermes is essentially the same as in the case of a single bad node depicted in Fig. 4.2. Nodes X_1 and X_2 misforward packets forwarded to them along this route with probabilities $B_f^{X_1}$ and $B_f^{X_2}$, respectively.

Node X_1's upstream neighbor node o_1 and X_2's upstream neighbor node o_2, initiate NACKs for all packets that are not acknowledged by nodes X_1 and X_2, respectively. Node Y_1 is able to verify the correct forwarding behavior of node o_1 at the MAC layer. Thus, upon receiving a NACK from node o_1, node Y_1 penalizes node X_1. However, nodes Y_1 and Y_2, upon receiving a NACK initiated by node o_2, penalize both nodes o_2 and X_2.

Fig. 4.4 illustrates the E-Hermes response to a a scenario of two bad neighbor nodes

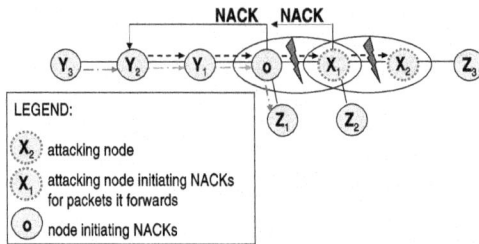

Figure 4.4: Multiple neighbor attacker nodes.

on a route. This attack scenario is similar to the previous one, with the difference being that the for flow f_1 along route $R_1 = \{Y_2, Y_1, o, X_1, X_2, Z_3, \cdots\}$, the bad nodes X_1, and X_2 are neighbors. Node X_1's upstream neighbor node o and X_2's upstream neighbor node X_1, following the acknowledgement scheme, will initiate NACKs for all packets that are not acknowledged by nodes X_1 and X_2, respectively. Node Y_2, upon receiving a NACK initiated by node o, penalizes both nodes o and X_1. Upon receiving a NACK initiated by node X_1, nodes Y_1 and Y_2 penalize both nodes X_1 and X_2.

Bad Recommender Nodes

The E-Hermes scheme relies on the exchange of trustworthiness information among nodes through recommendations. Thus, an obvious attack on the E-Hermes scheme would be for a given node to propagate false trustworthiness information, i.e., the node propagates a trustworthiness value that is different from the value that it should compute if it were following the Hermes scheme. Thus, a node may propagate a trustworthiness value that is higher or lower than the value that a Hermes-compliant node would compute.

The RC-test (see Section 4.4.1) ensures that recommendations are accepted only when the recommended trustworthiness value is close to the first-hand trustworthiness value that exceeds T_{acc} of the node that asked for the recommendations. If the first-hand trustworthiness value is smaller than T_{acc}, the node only temporarily accepts the maximum value from among all the recommenders. Because of this, bad recommender nodes are identified correctly, as also verified by our performance analysis in Section 4.7.

In this Section, we discuss the case of false categorization of a recommender node as a bad recommender, which does not influence the correct evaluation of the nodes as good or bad nodes. A bad recommender false positive only results in discarding the recommendations received by the recommender node. Fig. 4.5 summarizes the false positive bad recommender scenario. Here, source node Y_2 establishes route $R_1 = \{Y_2, Y_1, o, X, Z_2, \cdots\}$ for its flow f_1. Node X forwards data packets incorrectly with probability $0 < B_f^X \leq 1$. X's upstream neighbor node o will initialize NACKs for all packets that are not acknowledged by node X. Node Y_2 will penalize both nodes o and X.

Now assume that node Y_3 establishes route $R_2 = \{Y_3, Y_1, o, Z_1, \cdots\}$ for its flow f_2. Node Y_3

Y_3

NACK

Y_2 → Y_1 → o — X — Z_2

LEGEND:

X : attacking node

o : node initiating NACKs

Z_1

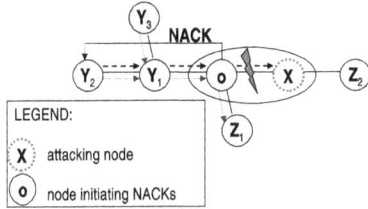

Figure 4.5: Bad recommender false positive.

sees node o as a good node. If nodes Y_2 and Y_3 exchange recommendations about node o at this point in time, they will consider each other as bad recommenders. However, as explained earlier, eventually node Y_2 will collect more network observations and will learn that node o is a good node. Then, the exchange of recommendations between nodes Y_2 and Y_3 would result in these nodes trusting each other as good recommenders.

Bad Node Sending Fake NACKs

In our attacker model, we have assumed that when a node forwards packets correctly, it also propagates ACK and NACK packets correctly, whereas when a node incorrectly forwards data packets, it does not initiate NACK packets. Here, we analyze the response of E-Hermes to the following attack: An attacker node X drops or misroutes data packets, and initiates NACK packets in an attempt to accuse its downstream neighbor of incorrect data packet forwarding. However, such an attack grants no advantage to the attacker, since the recommendation scheme penalizes both ends of the link at fault. As a result, all non-neighbor upstream nodes of the attacker X on the route will penalize the attacker node X and its downstream neighbor Z, if the attacker node initiates NACK packets. Otherwise, if the attacker node does not initiate fake NACK packets, but its upstream neighbor o initiates valid NACK packets, all non-neighbor upstream nodes of node o on the route will penalize the attacker node's upstream neighbor o and the attacker node X. In any case, the attacker node will be penalized.

Collusion of Bad Node and Bad Recommender

Fig. 4.6 illustrates the response of E-Hermes to a collusion involving a bad node and a bad recommender (yet good node) on a route. Here, source node Y_2 establishes route $R = \{Y_2, Y_1, w, X, Z_1, \cdots\}$ for its flow f. Node X and its upstream neighbor node w collude to perform the following attack. Node X forwards data packets incorrectly with probability $0 < B_f^X \leq 1$, while node w propagates incorrect trust information for node X of trustworthiness value higher than the value that a Hermes-compliant node would compute in an attempt to persuade the upstream nodes on the route that attacker node X is a good node. The goal of this attack is for node X to drop or misroute packets undetected, since its upstream neighbor w propagates incorrect high trustworthiness value for it.

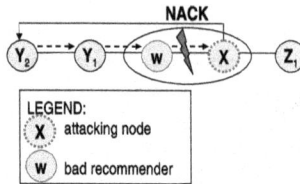

Figure 4.6: Attacker node and colluding bad recommender.

As part of the attack, node w does not initiate NACKs for the unacknowledged packets that node X incorrectly forwards, as otherwise it would be contradicting itself (NACKs from node w would indicate that link $w - X$ is faulty, which is the fact that w tries to hide by propagating incorrect trust information). Then, the attacker node X will send the NACKs itself, while dropping or misrouting packets. This is the case that we discussed in Section 4.6.3. In this case, all non-neighbor upstream nodes of the attacker X on the route will correctly penalize the attacker node X and its downstream neighbor Z_1. As discussed earlier, Z_1, will, with high probability, eventually be identified as a good node, when a more diverse set of observations data is accumulated in the network. Thus, node X's collusion with node w gives no benefits to node X, as the incorrect high trustworthiness value that node w propagated for it did not influence Hermes' ability to identify the misbehaving node correctly.

Wormhole Attack

If a wormhole attack occurs and no mechanism is in place to prevent this attack during the route discovery phase (cf. [79]), the E-Hermes scheme will still be able to identify the bad nodes involved in the wormhole. For example, suppose that two colluding nodes X and Y form a wormhole. They may be connected by a wired link or a wireless link formed using a directional antenna. Thus, nodes X and Y could form part of a route even though they are not neighbors. Now suppose packets are sent on a route which includes node X and Y. If the packets are not forwarded correctly by nodes X and Y through the wormhole, the acknowledgement scheme will penalize both nodes X and Y.

Sybil attack

In the Sybil attack, a node impersonates one or more of the other nodes in the network. This type of attack is an attack on the authentication scheme and can really only be addressed using cryptographic techniques (cf. [80]). Suppose a given node X launches a Sybil attack by impersonating another node Z on given route. Under the E-Hermes scheme, if node X drops packets sent along this route, then node Z will be penalized. On the other hand, node X is also penalized with respect to this route, since it claims to be node Z.

Attackers with directional antennas

Thus far, we have implicitly assumed that all nodes are equipped with omnidirectional antennas and that they do not employ dynamic power control. However, an outside adversary could use a directional antenna to launch an attack. E-Hermes can deal with this type of attack as follows. Suppose nodes X and Y are neighbors along a route and node Y has a directional antenna. We note that in order for node Y to be on the path, it must have compromised the authentication scheme of the network (since by assumption, all nodes in the network are equipped with only omnidirectional antennas). Now suppose node Y forwards packets in the direction of node X such that node X believes that node Y forwarded these packets correctly, when in actual fact, node Y does not forward these packets to the next node on the path. In this case, the first-hand trust evaluation based on neighborhood observations in Hermes will be foiled. However, the first-hand trust evaluation scheme based on acknowledgements will identify node Y correctly as the culprit. Thus, to deal with directional antenna attacks, the E-Hermes scheme should place more weight on the second form of first-hand trust or avoid the use of wireless channel snooping altogether.

4.7 Performance Evaluation of E-Hermes

In this Section, we evaluate the performance of Hermes. We first discuss the communication and computational overhead of our scheme. Then, we evaluate the accuracy of Hermes by presenting some representative results from our simulation experiments.

4.7.1 Communication and Computational Overhead

The E-Hermes scheme imposes both communication and computational overhead on the network. However, as discussed below, the overhead is reasonable, considering the security benefits provided by the scheme.

Communication Overhead

The communication overhead of E-Hermes consists of the following types of control packets:

- ACK and NACK packets transmitted as part of the acknowledgement scheme;

- Recommendation packets for propagating second-hand trust information.

These control packets can be piggybacked onto data packets whenever possible to reduce communication overhead. In any case, the control packets are relatively small, as discussed below.

For each data packet sent from the source node to the destination node of a given flow, an ACK or NACK packet must traverse the reverse path from the destination to the source node. Each ACK or NACK packet consist of the following three fields:

- type of packet;

- sequence number;

- authenticators.

Each authenticator is a hashed value consisting of 128 bits, if MD5 is used, or 160 bits, if SHA-1 is used (see Section 4.5.2). Consider a flow established on a route consisting of n hops, not including the source node. As an ACK packet traverses the reverse path from the destination node to the source node, a total of n authenticators is appended to the ACK packet. When a NACK is sent, k authenticators are appended to the NACK packet, where k is the number of nodes upstream from the node that initiated the NACK.

In the E-Hermes scheme, a recommendation request may be initiated by a node at the time a route is established (see Section 4.4.1). A recommendation request is sent to a set of recommender nodes chosen by the source node. Each recommendation request consists of the same fields listed above for ACK and NACK packets, except that only one authenticator is needed. Each recommendation reply packet has a similar format.

Computational Overhead

Each node maintains a set of values associated with each of the other nodes in the network:

- Counters A and M;

- Trust value t and confidence value c;

- Trustworthiness value T;

- Opinion value P;

- Recommender counters A^R and M^R;

- Recommender trustworthiness value T^R.

A given node updates a subset of the counters A and M each time it forwards a packet along a route to the next hop. For such a given packet, the counters A and M associated with the downstream nodes on the route are updated in accordance with the scheme for gathering first-hand trust information discussed in Section 4.3. The values t, c, T, and P are then updated whenever the A and M counters are updated. The values A^R, M^R, and T^R are updated whenever the RC-test is applied (see Section 4.4.1). To be responsive to changes in the network dynamics, the windowing mechanism presented in Section 3.4.4 should be applied in computing the trustworthiness and opinion values. For example, if a window of size K is used in computing the opinion value P, then the K most recent values of P over the averaging window must be stored. In summary, the storage and the computational requirements for maintaining the trustworthiness and opinion values in E-Hermes are relatively modest.

The acknowledgement scheme discussed in Section 4.5 incurs some additional computational overhead. The source node of a flow along a path of length n computes n MACs and constructs $2n - 1$ hash chains of length three, for each packet sent. Since hashing is a relatively inexpensive operation, the computational overhead is reasonable. Each intermediate node, upon receipt of an ACK or NACK packet, performs a hashing operation to compute an authenticator that is appended to the packet. The authentication of recommendation requests and replies, as discussed in Section 4.5, incurs little additional overhead since these messages are relatively short.

4.7.2 Performance Results

Simulation Methodology

We present some representative results from our simulation experiments for evaluating the accuracy of our scheme under different network and attack scenarios. The network consists of 10 nodes that are randomly placed in a 500 m by 500 m area. The wireless radio transmission range of the nodes is set to 250 m. Nodes exhibit four types of behavior.

- Type I: Good nodes and good recommenders;

- Type II: Bad nodes and good recommenders;

- Type III: Good nodes and bad recommenders;

- Type IV: Bad nodes and bad recommenders.

A predefined number of flows is generated for each simulation scenario. The route corresponding to a flow is not derived based on a given topology, but is chosen randomly to reflect the network topology at a given point in time. Thus, the effect of a dynamically changing network topology is captured in the simulation. The nodes in the network collect empirical evidence and build their trustworthiness and opinion values for all other network nodes based on traffic generated by the traffic flows.

Since the traffic flows are generated randomly, one or more misbehaving nodes may participate per flow. Misbehaving nodes may be neighbors or non-neighbors. The number of the traffic flows generated in the simulation scenarios presented in this Section is chosen to be small to highlight the fast and error-free convergence process of our scheme. However, when the number of generated flows is small, some nodes may not participate in any flows and as a result, no opinion is formed for them. Given a sufficiently large set of traffic flows, all nodes should be able to form valid opinions for every other node in the network. We remark that in the simulations discussed here, we do not employ the averaging windows introduced in Section 3.4.4, in order to simplify the presentation of results. Implementation of the averaging windows would have further improved the accuracy of the final opinions when the node behaviors change over time (see Fig. 4.9).

Network View

In the first simulation scenario, eight random traffic flows are established along different paths in the network. The minimum and maximum number of nodes allowed on a route are four and seven respectively. Nodes $1, 3, 4, 5, 8, 9, 10$ are randomly assigned to be of Type I. They forward 100% of the packets that they should be forwarding and propagate correct opinions P. Node 7 is randomly assigned to be of Type II. Node 7 forwards 20% of the packets received for forwarding, but propagates correct opinions P. Node 6 is randomly assigned to be of Type III. Node 6 forwards 100% of the packets received for forwarding, but propagates recommendations of fixed opinion $P = 0.5$. Node 2 is randomly chosen to be of Type IV. Node 2 forwards 20% of the packets received for forwarding, and propagates recommendations of fixed opinion $P = 0.5$. Although, in this case 30% of the nodes exhibit malicious behavior of one or another type, increasing this percentage does not affect the ability of the E-Hermes scheme to form accurate opinions. The source nodes send 100 data packets during each

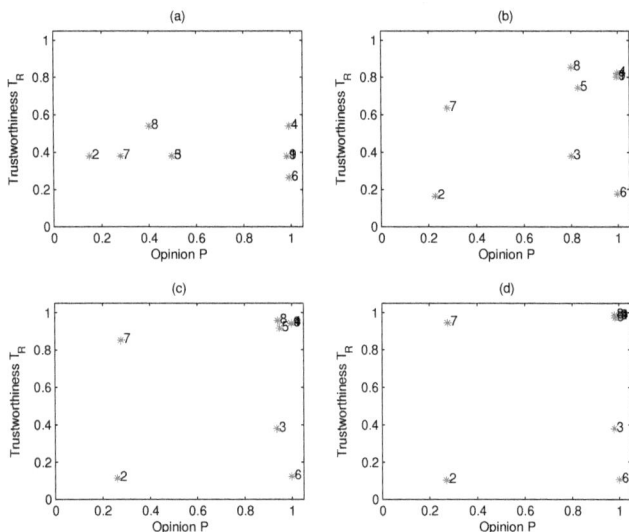

Figure 4.7: Opinion of good node/recommender for all other network nodes after (a) 1 round, (b) 3 rounds, (c) 10 rounds, (d) 30 rounds.

observation window W (also called "round"). The trustworthiness parameters are set as $x = \sqrt{2}$ and $y = \sqrt{9}$, and the RC-test threshold η is set to 0.1.

First, we implement our scheme with recommendations. Recommendations are exchanged among nodes that are in the same route and between any two nodes given that one of the nodes has formed opinions for nodes that the other node wants to use as intermediate nodes on a route. Figure 4.7 shows the opinion P and trustworthiness T^R that good node (and good recommender) 10 places on all other network nodes after 1, 3, 10, and 30 rounds. We note that node 10 identifies correctly the type of behavior of each node. Type I nodes appear in the top-right corner area, Type II nodes appear in the lower-right corner area, Type III nodes appear in the upper-left corner area, whereas Type IV nodes appear in the lower-left corner area. Node 3 is correctly identified as good node. However, node 10 has not formed recommender trustworthiness T^R for it, $T^R_{10,3} = T_{def}$, as node 10 had not asked node 3 for recommendations. Observe that the more observations node 10 makes, the more accurately it assigns opinion and recommender trustworthiness values.

Fig. 4.8 illustrates the opinion value that node i places on node j with a gray-scale representation. A black color implies an opinion value of 0, white represents an opinion value of 1, while intermediate values are represented by different shades of gray. Fig. 4.8 (b) illustrates the opinion values, $P_{i,j}$ which is the opinion formed in terms of packet forwarding. One can see that nodes 2 and 7 are identified as bad nodes by all other nodes, but node 6, which has

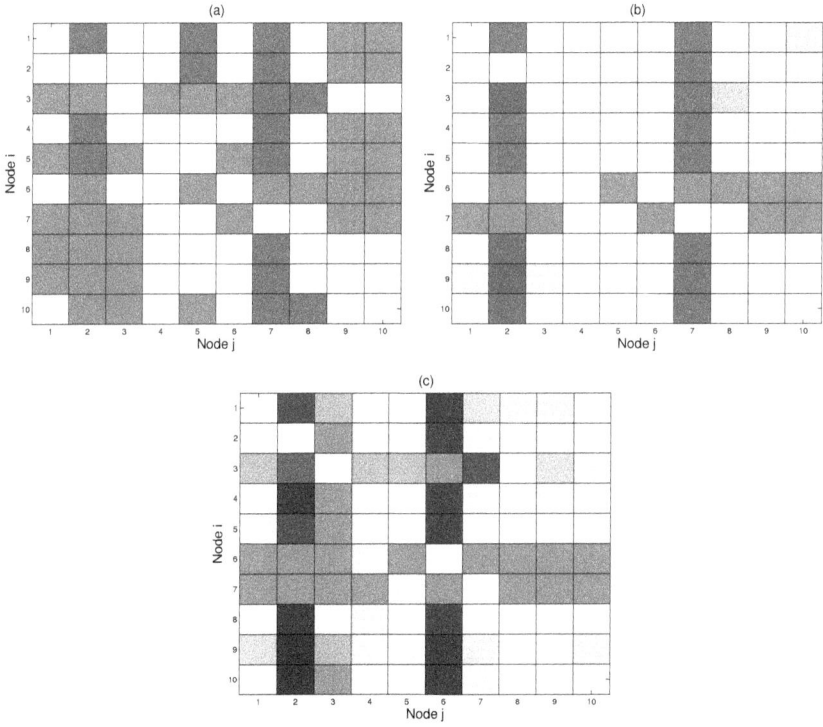

Figure 4.8: Network view: (a) Opinion $P_{i,j}$, without recommendations, (b) Opinion $P_{i,j}$, with recommendations, (c) Recommender trustworthiness $T_{Ri,j}$.

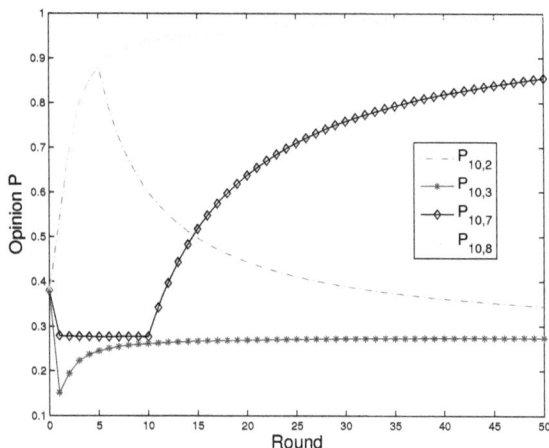

Figure 4.9: Opinion that node 10 forms for nodes $2, 3, 7, 8$ from round 1 to 50. Nodes $2, 7$ change their forwarding behavior in rounds 5 and 10 respectively.

not interacted with them and is ignorant about their behavior ($T_{6,2} = T_{6,7} = T_{def}$). Node 7 has not identified node 2 as bad node for the same reason. The good nodes are also correctly identified.

Fig. 4.8 (c) shows the recommender trustworthiness values, $T_{i,j}^R$, which are the opinions formed in terms of trust propagation. Nodes 2 and 6 are correctly identified as bad recommenders by all other nodes that were able to form acceptable recommender trustworthiness values T^R for them. The remaining nodes are correctly identified as good recommenders with one exception. There is a false positive recommender trustworthiness T^R, because only eight flows are active. With more flows in the network, the accuracy of the opinions formed improves. However, note that the existence of false positives T^R is acceptable, as long as the correct opinions P are formed, which is the case here.

Fig. 4.8 (a) illustrates the opinion values, $P_{i,j}$ when the scheme without recommendations is implemented for the same simulation scenario. Nodes that have interacted with nodes 2 and 7 have correctly identified them as bad nodes. Nodes that interacted with the remaining nodes have identified them as good, with two exceptions. There are two false positives that are attributed to the fact that upon a NACK receipt both nodes of the faulty link are suspected as bad nodes. This effect attenuates when many flows are established, given that bad nodes have different neighbors. Comparing (a) and (b) we see that when recommendations are used, nodes form the correct network view much more quickly. We have also tested our scheme under various attack scenarios, varying the number of bad recommenders and bad nodes, and found that the scheme forms accurate opinions in all cases.

Adaptive Behavior

To demonstrate our scheme's ability (with recommendations) to adapt to changes in the node behaviors, we use the same simulation scenario. Eight flows are generated and the source nodes send 100 data packets during each round. The simulation runs for fifty rounds. However, now nodes $1, 4, 5, 8, 9, 10$ are of Type I. Nodes $2, 6$ are bad recommenders, propagating opinions with value $P = 0.5$. Node 3 is of Type II. Node 2 is good for rounds 1-5 and then becomes bad, thus switching from Type III to Type IV. Node 7 is bad for rounds 1-10 and then becomes good, thus switching from Type II to Type I. Node 6 is of Type III. Good nodes forward 100% of the packets that they should be forwarding. Bad nodes forward 20% of the packets received for forwarding. As before, the RC-test threshold η is set to 0.1.

The opinions P that node 10 places on nodes $2, 3, 7, 8$ over 50 rounds is shown in Fig. 4.9. Our scheme accurately evaluates trust and adapts to changes in the nodes' behaviors. Note that the past behavior of a node influences the value of the current opinion P. For example, at round 50 $P_{10,8} \approx 1$, whereas $P_{10,7} = 0.86$. The implementation of the windowing mechanisms as proposed in Section 3.4.4 would systematically expire old observation data in order to improve the responsiveness of the system. We remark that the ability of Hermes scheme to quickly adapt to changing node behavior is a key point of the scheme that makes it practical for real-world networks.

Convergence Comparison

In this simulation, we compare the convergence of our scheme with and without the use of recommendations. The objective is to investigate the BN-recognition of our scheme as a function of active network flows. The simulated network consists of 10 nodes. Nodes $1, 3, 4, 5, 8, 9, 10$ are of Type I, node 7 is of Type II, node 6 is of Type III and node 2 of Type IV. As in earlier simulations, good nodes forward 100% of packets, bad nodes 20%, good recommenders propagate valid trust values, whereas bad recommenders send $P = 0.5$. Initially one flow is generated and then one flow is added per round. The flows are randomly generated. The number of nodes on a route is set to 5.

Figure 4.10 shows the BN-recognition of the scheme with and without recommendations. The error bars indicate the 90% confidence intervals obtained from executing on the order of 20 simulation trials for each estimated value. As expected, recommendations accelerate the convergence of the trust establishment procedures. With recommendations, the BN-recognition converges to a steady-state value after fewer than 20 rounds, whereas when recommendations are not used, more than 35 rounds are required for the scheme to converge.

Misbehavior Recognition

A useful measure of the performance of the proposed trust establishment scheme is given as follows.

Definition 13. *The **misbehavior(ϵ) recognition** percentage or* MB(ϵ)-recognition *is the percentage of the nodes in the network that have identified all the misbehaving nodes in the network by forming the correct opinions $P_{i,m} = 1 - B_f^i$ for them with precision of ϵ, i.e.,* $|P_{i,m} - (1 - B_f^i)| < \epsilon$.

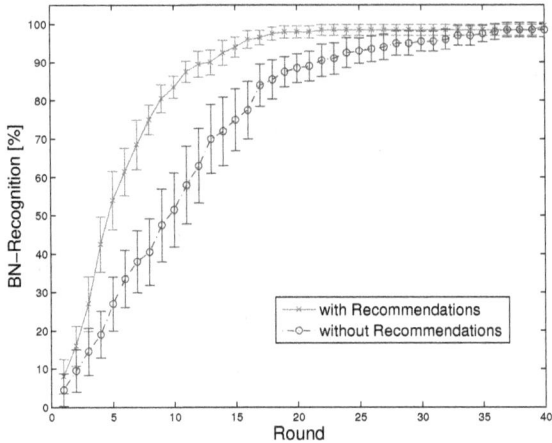

Figure 4.10: Convergence comparison of scheme with and without recommendations in respect to BN-recognition.

First, we run a number of simulations to show 1) the MB(0.1)-recognition and 2) the convergence rate of E-Hermes when the percentage of misbehaving network nodes and the probability of incorrect packet forwarding behavior B_f of the misbehaving nodes changes. In particular, the percentage of misbehaving network nodes ranges from 4% to 95%, whereas the probability of incorrect packet forwarding behavior B_f ranges from 20% to 100%. The number of bad recommenders is set to 25% of the network nodes. Then, we run a number of simulations to show 1) the MB(0.1)-recognition and 2) the convergence rate of E-Hermes when the percentage of bad recommenders and the probability of incorrect packet forwarding behavior B_f of the misbehaving nodes changes. In particular, the percentage of bad recommenders ranges from 4% to 100%, whereas the probability of incorrect packet forwarding behavior B_f ranges from 20% to 100%. The number of misbehaving nodes is set to 25% of the network nodes.

The misbehaving nodes and the bad recommenders are chosen randomly from the set of the network nodes. Thus, a node may exhibit any of the four Types of behavior introduced in Section 4.7.2. The simulated network consists of 30 nodes. Initially one flow is generated and then one flow is added per round. The flows are randomly generated. The number of nodes on a route is set to 7. The non-misbehaving nodes forward all the packets that they receive for forwarding. The good recommenders propagate correct opinions P. The bad recommenders propagate fixed opinion $P = 0.5$ when the probability of incorrect packet forwarding behavior $B_f \neq 0.5$ and they propagate fixed opinion $P = 0.2$ when the probability of incorrect packet forwarding behavior $B_f = 0.5$. The source nodes send 100 data packets during each observation window W (also called "round"). The trustworthiness parameters

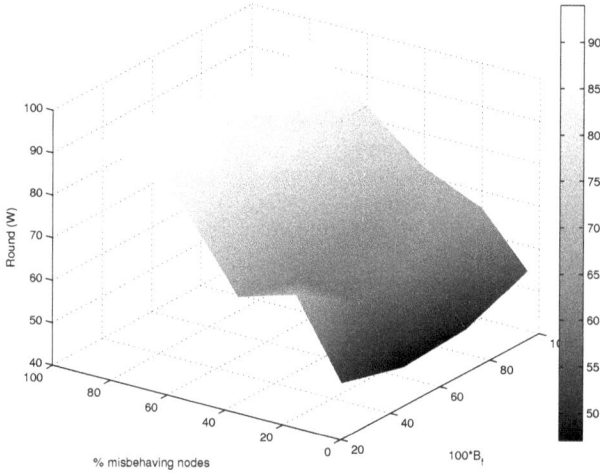

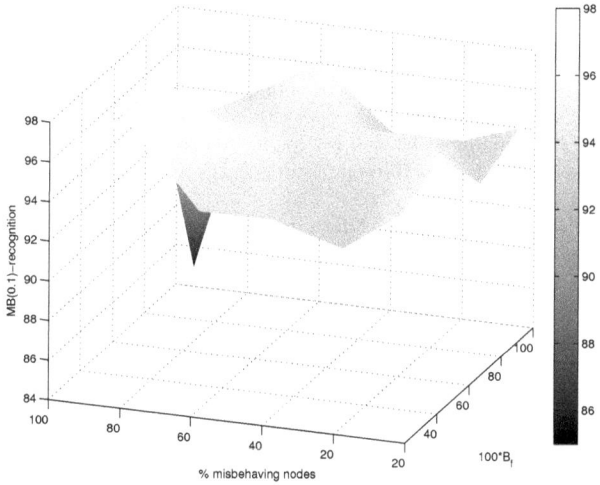

Figure 4.11: (a) Convergence rate, (b) MB(0.1)-recognition, when the percentage of misbehaving network nodes and the probability of incorrect packet forwarding behavior B_f of the misbehaving nodes changes.

are set as $x = \sqrt{2}$ and $y = \sqrt{9}$, and the RC-test threshold η is set to 0.1. The results are obtained from executing on the order of 10 simulation trials for each network scenario.

Figure 4.11 (a) illustrates the number of rounds (or windows W) required for E-Hermes to reach a steady state. As expected, the convergence rate depends on the percentage of misbehaving network nodes. The more the misbehaving nodes in the network, the longer it takes for E-Hermes to reach a steady state. The probability of incorrect packet forwarding behavior B_f of the misbehaving nodes slightly influences the convergence rate of the proposed trust establishment framework. For example, when there are 4% of misbehaving network nodes, E-Hermes requires $50, 47, 48$, and 54 rounds to reach steady state when B_f equals to $0.25, 0.5, 0.75$ and 1.0 respectively. When there are 95% of misbehaving network nodes, E-Hermes requires $89, 90, 86$, and 76 rounds to reach steady state when B_f equals to $0.25, 0.5, 0.75$ and 1.0 respectively.

Figure 4.11 (b) shows the MB(0.1)-recognition of E-Hermes on steady state. As expected, the MB(0.1)-recognition of E-Hermes is on the range of $93 - 98\%$, with one exception. Only when there are 95% of misbehaving network nodes with $B_f = 1.0$, the MB(0.1)-recognition of E-Hermes is 85%. Nonetheless, it should be noted that the steady state is reached only after 76 rounds.

Figure 4.12 (a) illustrates the number of rounds (or windows W) required for E-Hermes to reach a steady state. As expected, the convergence rate depends on the percentage of bad recommenders in the network. However, the number of bad recommenders in the network, does not influence the convergence rate in a linear fashion. For example, when there are $4 - 50\%$ of bad recommenders in the network, and the misbehaving nodes exhibit incorrect packet forwarding behavior $B_f = 0.75$, E-Hermes requires $72 - 74$ rounds to reach steady state respectively. When the percentage of bad recommenders in the network is increased from 75 to 90, and 100%, and the misbehaving nodes exhibit incorrect packet forwarding behavior $B_f = 0.75$, E-Hermes requires $80, 96$, and 182 rounds to reach steady state respectively. It can be concluded that even 10% of good recommenders in the network can significantly accelerate the convergence rate of E-Hermes.

The probability of incorrect packet forwarding behavior B_f of the misbehaving nodes slightly influences the convergence rate of the proposed trust establishment framework. For example, when there are 75% of bad recommenders in the network, E-Hermes requires $83, 86, 80$, and 84 rounds to reach steady state when B_f equals to $0.25, 0.5, 0.75$ and 1.0 respectively. Nonetheless, we can notice a decreased convergence rate. When there are 100% of bad recommenders in the network, and the misbehaving nodes exhibit incorrect packet forwarding behavior $B_f = 1.0$, E-Hermes requires 230 rounds to reach steady state.

Figure 4.12 (b) shows the MB(0.1)-recognition of E-Hermes on steady state. As expected, the MB(0.1)-recognition of E-Hermes is on the range of $93 - 99\%$, with one exception. Only when there are 100% of bad recommenders in the network, and the misbehaving nodes exhibit incorrect packet forwarding behavior $B_f = 1.0$, the MB(0.1)-recognition of E-Hermes is 84%.

Figure 4.13 (a) compares the number of rounds (or windows W) required for E-Hermes to reach a steady state when 1) 100% of the nodes in the network are bad recommenders, and 2) recommendations are not exchanged in the network. Figure 4.13 (b) shows the MB(0.1)-recognition of E-Hermes on steady state when 1) 100% of the nodes in the network are bad recommenders, and 2) recommendations are not exchanged in the network. The objective is to show that the exchange of bad recommendations does not undermine the performance of E-Hermes. This can be verified by the presented results. Recommendations are used to

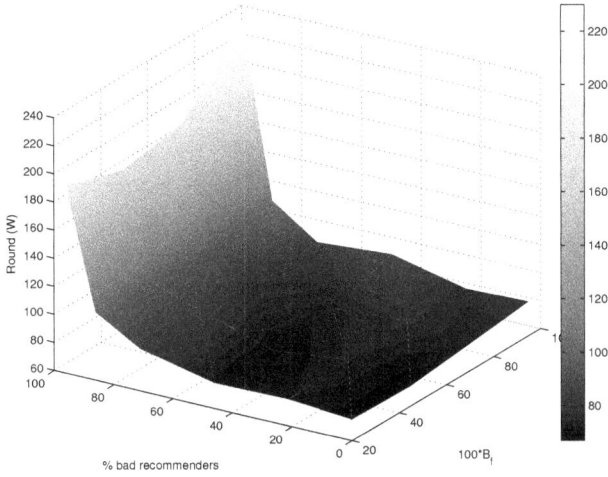

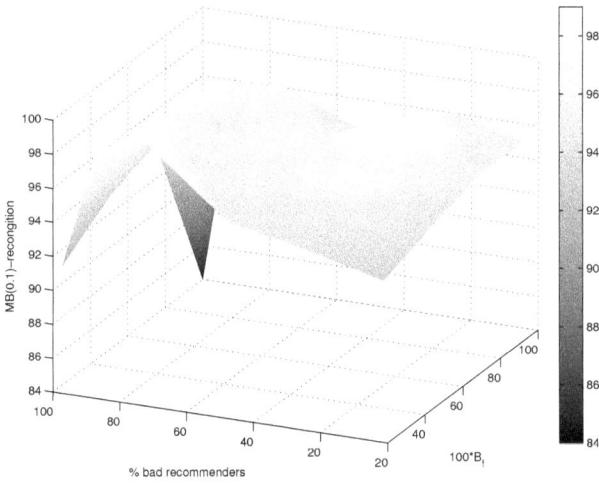

Figure 4.12: (a) Convergence rate, (b) MB(0.1)-recognition, when the percentage of bad recommenders and the probability of incorrect packet forwarding behavior B_f of the misbehaving nodes changes.

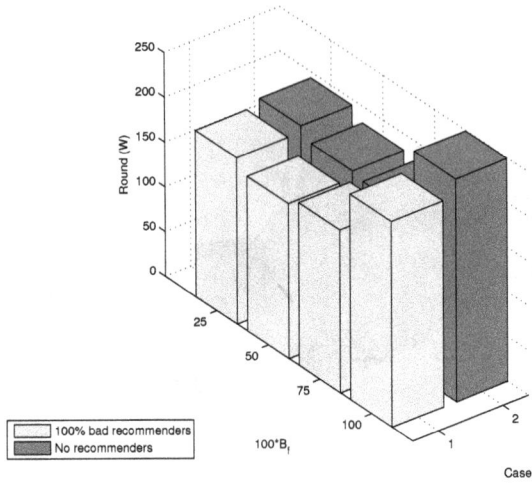

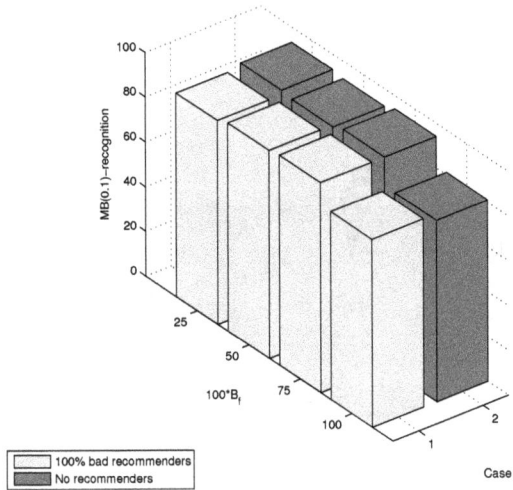

Figure 4.13: (a) Convergence rate, (b) MB(0.1)-recognition, when 1) 100% of the nodes in the network are bad recommenders, and 2) recommendations are not exchanged in the network.

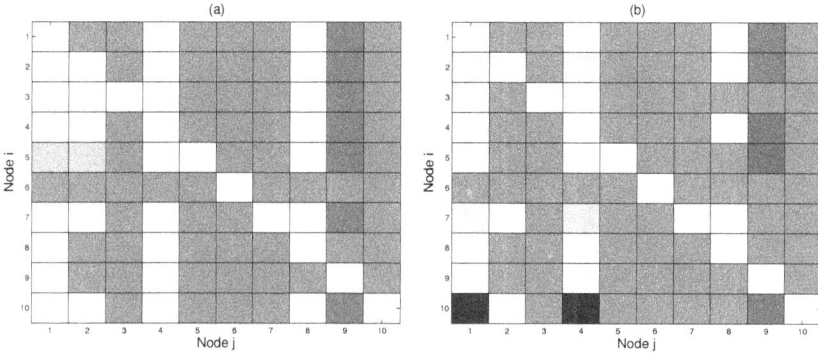

Figure 4.14: Network view: (a) Opinion $P_{i,j}$ of E-Hermes, (b) Opinion $P_{i,j}$ of Hermes.

accelerate the convergence of the trust establishment procedures. If all the nodes in the network are bad recommenders, E-Hermes performs as if no recommenders are present in the network.

E-Hermes vs. Hermes

Fig. 4.14 compares the performance of the E-Hermes scheme with the original Hermes scheme in a gray-scale representation. The comparison is done for the simulation scenario presented in Section 3.7.3. Five traffic flows are established in a network of ten nodes. Node 9 acts maliciously, forwarding only 20% of the packets. Hermes assumes that malicious nodes are also bad recommenders. The trust values they propagate are ignored and T_{def} is used for the trustworthiness calculation of nodes downstream of a malicious node. We simulated a scenario in which node 9 is a bad recommender that propagates $P = T_{def}$. All other nodes forward 90% of the packets they should be forwarding. The source node of each flow sends 20 packets per window over the course of 30 rounds.

Figure 4.14 (a) illustrates the opinion values $P_{i,j}$, that node i places on node j when our scheme is implemented and Fig. 4.14 (b) is the equivalent network view presented in Section 3.7.3. One can verify that the E-Hermes scheme is superior in terms of (i) convergence time and (ii) accuracy of the opinion values in the sense that more nodes correctly identify other nodes as good or bad. In Fig. 4.14 (a), we see that seven nodes have identified node 9 as bad, whereas in Fig. 4.14 (b) only five nodes have identified node 9 as bad.

Large Network Scenario

We now simulate a larger network scenario consisting of 100 nodes and 200 random traffic flows established along different paths in the network. The minimum and maximum number of nodes allowed on a route are four and ten, respectively. We chose 2 flows to be initiated in average per node to emphasize the convergence of our scheme with a limited number of traffic

74

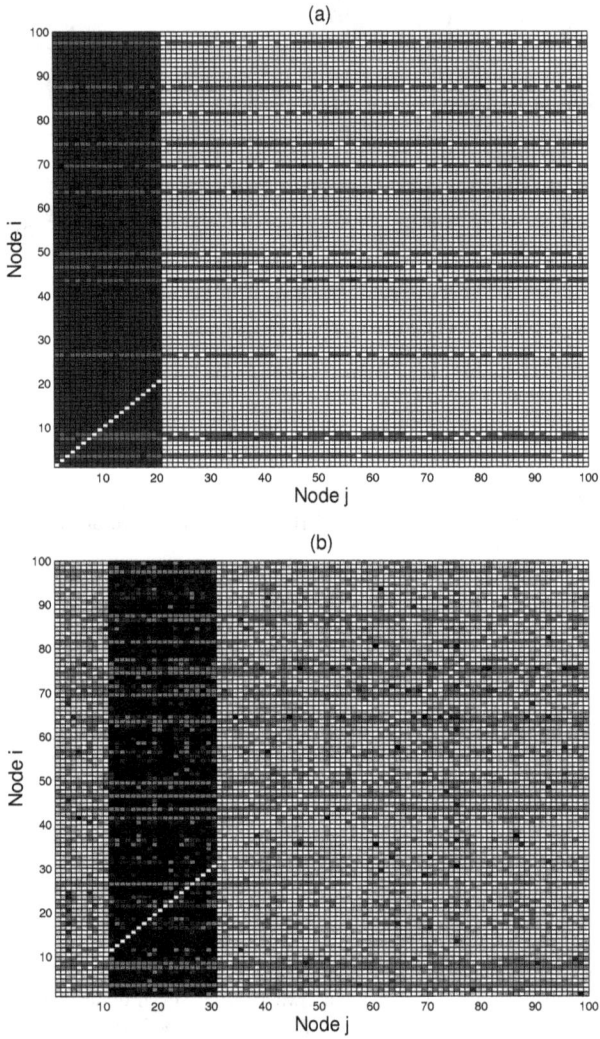

Figure 4.15: Large-scale network view: (a) Opinions $P_{i,j}$, (b) Recommender trustworthiness values $T_{i,j}^R$.

flows. Nodes 1-10 are assigned to be of Type II and forward 20% of the packets received for forwarding, but propagate correct opinions P. Nodes 21-30 are assigned to be of Type III; they forward 100% of the packets received for forwarding, but propagate recommendations of fixed opinion $P = 0.5$. Nodes 11-20 are chosen to be of Type IV; they forward 20% of the packets received for forwarding, and propagate recommendations of fixed opinion $P = 0.5$. The remaining nodes are of Type I; they forward 100% of the packets that they should be forwarding and propagate correct opinions P. Thus, 30% of the network nodes exhibit malicious behavior of one or another type. The source nodes send 50 data packets during each observation window W. The trustworthiness parameters are set as before. Recommendations are used in the simulation.

Figure 4.15 (a) illustrates the opinion values $P_{i,j}$, that node i places on node j. One can see that nodes 1-20 are identified as bad nodes and nodes 21-100 are identified as good nodes by 87% of nodes; 13% nodes did not form an opinion about one or more of nodes, because of lack of interaction with them. In total, $9,900$ opinions are formed. The number of false positives was 16, which corresponds to 0.0016% of all opinions. The false positives are attributed to the fact that upon a receipt of a NACK both nodes of the faulty link are suspected as bad nodes. As mentioned earlier, this effect is attenuated by the presence of a larger number of diverse flows which contain bad nodes with a variety of good neighbors. These results suggest the effectiveness of E-Hermes in larger network scenarios.

Figure 4.15 (b) shows the recommender trustworthiness values, $T_{i,j}^R$, that node i places on nodes j. One sees that nodes 11-30 are correctly identified as bad recommenders by all other nodes that were able to form acceptable recommender trustworthiness values for them. The remaining nodes are correctly identified as good recommenders by the majority of the nodes. We note that there are some false positives in the recommender trustworthiness values. However, the existence of false positives T^R is acceptable as long as the correct opinions P are formed, which is the case here.

E-Hermes for Distance Vector Routing

In the previous simulation experiments we assumed that E-Hermes was implemented on top of a source routing algorithm. As a result, our proposed acknowledgement algorithm was simulated and nodes formed opinions for non-neighbor nodes. As discussed in Section 4.3.2, for distance vector routing schemes, the acknowledgement scheme is not needed since nodes only form opinions for their neighbor nodes. In this Section, E-Hermes is assumed to be implemented on top of a distance vector routing algorithm.

Fig. 4.16 illustrates the opinion value that node i places on node j with a gray-scale representation. The scenario of Section 4.7.2 is simulated. Fig. 4.16 (a) illustrates the opinion values, $P_{i,j}$ when the E-Hermes scheme without recommendations is implemented. Nodes that have interacted with nodes 2 and 7 have correctly identified them as bad nodes. Nodes that interacted with the remaining nodes have identified them as good. Each node is responsible for evaluating trust for its neighbor nodes. If a neighbor node is identified as bad node, another neighbor node is chosen for packet forwarding. For non-neighbor nodes the trust established is T_{def}. Fig. 4.16 (b) illustrates the opinion values, $P_{i,j}$ for our scheme with recommendations. Recommendations are exchanged so that when nodes move and their neighbor set changes, they have an initial trust established on them by their recommenders. We note that even though nodes 2 and 6 are bad recommenders no wrong opinions are formulated.

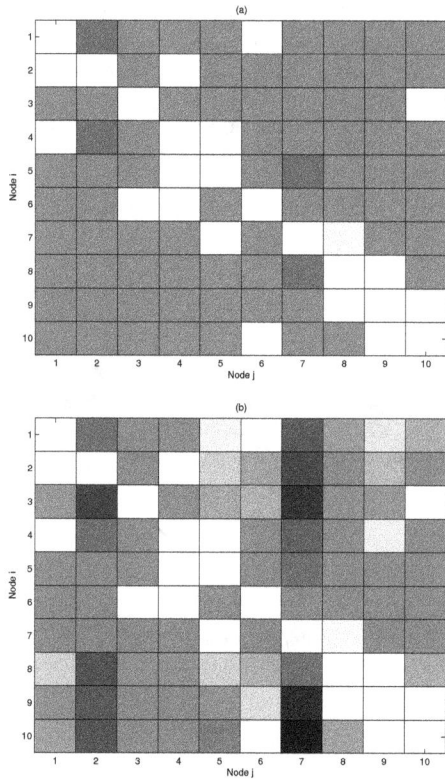

Figure 4.16: Network view for distance vector routing (a) Opinion $P_{i,j}$ without Recommendations, (b) Opinion $P_{i,j}$ with Recommendations.

4.8 E-Hermes Summary

We presented a robust cooperative trust establishment scheme for MANETs, which is designed to improve the reliability of packet forwarding over multi-hop routes, particularly in the presence of malicious nodes. The proposed scheme extends the Hermes framework of Chapter 3 in several important ways. In the E-Hermes scheme, first-hand information for non-neighbor nodes is obtained via feedback from acknowledgements sent in response to data packets. The E-Hermes exploits information sharing among nodes to accelerate the convergence of trust establishment procedures. Second-hand trust information is obtained via recommendations from cooperative nodes. The trustworthiness of the recommendations and recommenders is evaluated. The concept of trustworthiness is then extended to the notion of an *opinion* that a given node has about the forwarding behavior of any arbitrary node by combining first-hand and second-hand trust information.

The proposed extensions to Hermes allow nodes to form accurate opinions for any network node and provides robustness against the propagation of false trust information by malicious nodes. The number of nodes that propagate false trust information does not influence the robustness of the system. Three types of malicious node behavior are identified: (i) dropping or misrouting packets but propagating true opinion values, (ii) forwarding packets but propagating false opinion values, and (iii) dropping or misrouting packets and propagating false opinion values. The effect of attacks by malicious nodes is identified either when they operate separately or form collusions. We presented simulation results which demonstrate the effectiveness of the E-Hermes scheme in distinguishing among malicious and non-malicious nodes in a variety of network scenarios involving nodes that are malicious both with respect to packet forwarding and trust propagation.

Chapter 5: Byzantine Robustness

In this Chapter, we propose new mechanisms to make Hermes robust to Byzantine behavior, i.e., arbitrary, deviant behavior, which disrupts packet transmission in the network and introduce a punishment policy that discourages selfish node behavior. Most of the material in this Chapter also appears in [81].

5.1   Overview

A deficiency of the E-Hermes scheme is that a node can still fail to detect certain types of Byzantine behavior, i.e., behavior that can deviate in an arbitrary manner [61, 82]. In particular, the E-Hermes scheme is vulnerable to attacks by nodes that misforward data packets in a selective manner. For example, a Byzantine node may choose to drop packets belonging to a certain set of source nodes, or to forward only packets belonging to another set of source nodes. Consider the example scenario shown in Fig. 5.1. Let

$$R_1 = \{i, a_1, ..., a_j = j, a_{j+1} = B, ..., a_{n-1}, a_n = m\},$$

denote a path from node i to node m, and

$$R_2 = \{j, b_1 = B, b_2, ..., b_{n-1}, b_n = m'\},$$

denote a path from node j to node m'. Note that node j is an intermediate node on the path R_1 and the source node of the traffic flow that traverses path R_2. Suppose that node B exhibits Byzantine behavior by *correctly* forwarding all the packets that node i sends to it on path R_1, but *incorrectly* forwarding 20% of the packets that node j sends to it on path R_2. Assume that all other nodes forward all packets *correctly*.

If both source nodes i and j send 100 packets for forwarding to their respective destinations, node j observes that node B *correctly* forwards 120 of the 200 packets that it received for forwarding. Node j's goal is to determine which nodes it should "trust" for forwarding its

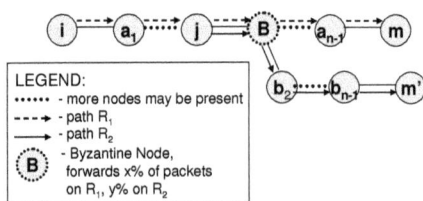

Figure 5.1: Example scenario illustrating the Byzantine node problem.

data packets to their destination nodes. If node j does not differentiate between the empirical evidence that it accumulates as a source from the empirical evidence that it accumulates as intermediate node, it will not be able to identify which nodes exhibit Byzantine behavior for its flows. Under the E-Hermes framework, node j only observes that node B forwarded 120 out of 200 packets. Hence, node j is unable to recognize that node B is exhibiting Byzantine behavior. This suggests that node j should rely mainly on observations of node forwarding behavior with respect to its own packets, i.e., 20 out of the 100 packets it forwards in this example.

While the E-Hermes scheme provides a means for a node to determine the trustworthiness of other nodes, it does nothing to discourage nodes from acting selfishly. For example, a node could simply drop all packets forwarded to it in order to conserve its own battery power. Under E-Hermes, such a node would be identified as a "bad" node by other nodes, but it would have no incentive to alter its behavior if its primary objective were to conserve battery power. Thus, a trust establishment scheme alone is not sufficient to alleviate the effects of Byzantine node behavior.

In this Chapter, we propose new mechanisms that make the Hermes scheme robust to Byzantine behavior. We also introduce a punishment policy that discourages selfish node behavior. We refer to the new trust establishment scheme, including the punishment policy, as *Byzantine Robust Hermes* or *BR-Hermes*, to distinguish it from the earlier Hermes and E-Hermes schemes mentioned above. The main difference between BR-Hermes and E-Hermes is that in the BR-Hermes scheme, each node distinguishes whether it is a source node or intermediate node with respect to a given packet flow that provides first-hand information on node behavior. This information is then used to compute a first-hand trust metric that is robust to Byzantine node behavior. An additional benefit of the BR-Hermes scheme is that its property to identify Byzantine behavior can be exploited to develop a punishment policy that discourages nodes from selfishly dropping packets. The main features provided by the BR-Hermes scheme are summarized as follows:

- ability to determine accurate trustworthiness information for any node in the network in the presence of Byzantine nodes;

- responsiveness to changes in node behavior;

- ability to distinguish between malicious behavior with respect to packet forwarding vs. trust propagation;

- ability to identify the effect of attacks by individual or colluding malicious nodes.

The remainder of the Chapter is organized as follows. Sections 5.2 and 5.3 discuss the core concepts and advances of the Chapter. Section 5.2 presents a Byzantine robust scheme for accumulating trust information and computing trust metrics for neighbor and non-neighbor nodes. Section 5.3 introduces a punishment scheme that discourages selfish node behavior. In Section 5.4, we present a security evaluation of our BR-Hermes trust establishment scheme. Section 5.5 presents results from simulation experiments that demonstrate the performance properties of BR-Hermes. In particular, we provide numerical comparisons of the BR-Hermes scheme versus the E-Hermes scheme, which does not detect Byzantine behavior. The overhead incurred by BR-Hermes is also discussed in Section 5.5. Finally, the Chapter is concluded in Section 5.6.

5.2 Trust Evaluation for Byzantine Detection

In this section, we present the BR-Hermes scheme for gathering first-hand trust information from neighbor and non-neighbor nodes. The first-hand information that a node obtains from a traffic flow is weighted depending on whether the given node is the source or an intermediate node on the route that the flow traverses. This is an extension over Hermes and E-Hermes, which gather first-hand information without considering the role of the node on the route. The BR-Hermes scheme requires authentication mechanisms for both data and control packets. The authentication mechanisms discussed in Section 4.5 can be applied here.

5.2.1 Problem Statement

Consider a very simple route $\{x, y, z\}$. In E-Hermes scheme, a given node x in the network maintains counters M_y and A_y for node y. We refer to the sets of counters $\{M_y\}$ and $\{A_y\}$ as M-counters and A-counters, respectively. The counter M_y records the total number of packets sent from node x to node y for forwarding to z over an observation window. The counter A_y records the total number of packets forwarded *correctly* (not dropped or misrouted) from node y to node z. Then the trust and confidence that x attributes to y over an observation window[1] are given by (cf. (4.1) and (4.2))

$$t_{x,y} = t(A_y, M_y) \text{ and } c_{x,y} = c(A_y, M_y),$$

from which the trustworthiness value $T_{x,y}$ can be computed via (4.3).

With respect to the scenario of Fig. 5.1, we discussed the significance of the empirical evidence that a node collects as a source node in a network. However, this should not minimize the importance of the empirical evidence that can be accumulated when a node serves as an intermediate node on paths in the network. To better understand this point, reconsider the scenario of Fig. 5.1, with the difference that now node B forwards *correctly* all the packets that node i sends to it on path R_1, and all the packets that node j sends to it on path R_2. If both source nodes i and j send 100 packets for forwarding, node j observes that node B forwards *correctly* all 200 packets that it received for forwarding.

Taking these remarks into consideration, we proceed to develop the BR-Hermes scheme, which distinguishes the empirical evidence collected as a source and the empirical evidence collected as an intermediate node and combines them to compute a Byzantine robust trust metric.

5.2.2 Byzantine Robust Processing of Empirical Evidence

In this section, we present a new scheme for processing the first-hand trust information from neighbor and non-neighbor nodes. The key point of our new scheme is that a node differentiates between the empirical evidence that it collects when it is the source of traffic flows and the empirical evidence that it collects when it is an intermediate node on routes that flows of other sources traverse.

---

[1]Windowing methods to expire old observation data are discussed in section 3.4.4.

Weighting Empirical Evidence

Let $R = \{a_0, a_1, a_2, \cdots, a_{n-1}, a_n\}$, where $n \geq 2$, denote a path from node a_0 to node a_n. In the BR-Hermes scheme, the source node a_0 maintains counters

$$\widetilde{M}_{a_i} \quad \text{and} \quad \widetilde{A}_{a_i}, \quad 1 \leq i \leq n-1.$$

for its downstream nodes. The intermediate nodes a_j, $1 \leq j \leq n-2$ maintain counters

$$\widehat{M}_{a_i} \quad \text{and} \quad \widehat{A}_{a_i}, \quad j+1 \leq i \leq n-1.$$

for their downstream nodes.

We refer to the sets of counters $\{\widetilde{M}_{a_i}\}$ and $\{\widetilde{A}_{a_i}\}$, $1 \leq i \leq n-1$ as $\widetilde{M}$-counters and $\widetilde{A}$-counters respectively. Similarly, we refer to the sets of counters $\{\widehat{M}_{a_i}\}$ and $\{\widehat{A}_{a_i}\}$, $1 \leq j \leq n-2$, $j+1 \leq i \leq n-1$, as $\widehat{M}$-counters and $\widehat{A}$-counters respectively. The counters M record the total number of packets sent for forwarding over an observation window. The counters A record the total number of packets forwarded *correctly* (i.e., not dropped or misrouted) from the downstream nodes. The number of packets forwarded *incorrectly* by the downstream nodes is given by $B \triangleq M - A$.

- $\widetilde{M}$-counters and $\widetilde{A}$-counters indicate the empirical evidence collected from the network when the node is the source of the traffic flow;

- $\widehat{M}$-counters and $\widehat{A}$-counters indicate the empirical evidence collected from the network when the node is an intermediate node on the path that the traffic flow traverses.

Then the trust $\widetilde{t}_{a_0, a_i}$ and confidence $\widetilde{c}_{a_0, a_i}$ that the source node a_0 attributes to its downstream nodes a_i, $1 \leq i \leq n-1$ over an observation window are given by (cf. (4.1) and (4.2))

$$\widetilde{t}_{a_0, a_i} = t(\widetilde{A}_{a_i}, \widetilde{M}_{a_i}), \quad 1 \leq i \leq n-1$$

and

$$\widetilde{c}_{a_0, a_i} = c(\widetilde{A}_{a_i}, \widetilde{M}_{a_i}), \quad 1 \leq i \leq n-1$$

from which the trustworthiness value $\widetilde{T}_{a_0, a_i}$, $1 \leq i \leq n-1$ can be computed via (4.3).

Similarly, the trust $\widehat{t}_{a_j, a_i}$ and confidence $\widehat{c}_{a_j, a_i}$ that intermediate node a_j, $1 \leq j \leq n-2$, attributes to its downstream nodes a_i, $j+1 \leq i \leq n-1$ over an observation window are given by (cf. (4.1) and (4.2))

$$\widehat{t}_{a_j, a_i} = t(\widehat{A}_{a_i}, \widehat{M}_{a_i}), \quad 1 \leq j \leq n-2, \quad j+1 \leq i \leq n-1$$

and

$$\widehat{c}_{a_j, a_i} = c(\widehat{A}_{a_i}, \widehat{M}_{a_i}), \quad 1 \leq j \leq n-2, \quad j+1 \leq i \leq n-1$$

from which the trustworthiness value $\widehat{T}_{a_j, a_i}$, $1 \leq j \leq n-2$, $j+1 \leq i \leq n-1$ can be computed via (4.3).

The next step is to define a method, which will allow every node to take advantage of the empirical evidence that it collects as intermediate node, without risking to lose its visibility of the network nodes' forwarding behavior for its own traffic flows. In other words, we wish to assign a weight w to the empirical evidence that a nodes collects as an intermediate node, where $0 \leq w \leq 1$. A value of w close to one indicates that the given evidence weights by 100%, whereas a value close to zero indicates that the given evidence does not weight at all, and thus is ignored. When the node is the source of a flow, the information it collects from the network is weighted by $w = 1$.

To decide on the weight w that will be assigned to the information collected by a node when it is an intermediate node on the route that the flow traverses, we calculate the absolute value of the difference between the trustworthiness value $\widetilde{T}_{i,m}$ that node i forms for a node m and the trustworthiness value $\widehat{T}_{i,m}$ that node i forms for a node m:

$$\text{T-Distance} : |\widetilde{T}_{i,m} - \widehat{T}_{i,m}| = e \tag{5.1}$$

where $e \in [0,1]$, since $\widetilde{T}_{i,m}, \widehat{T}_{i,m} \in [0,1]$. Then, the weight w corresponding to the counters $\widehat{A}$ and $\widehat{B}$ is assigned as

$$w = 1 - e. \tag{5.2}$$

The smaller the value of the T-Distance, the more node i comes to believe that node m does not exhibit Byzantine behavior, since it behaves similarly when it forwards data packets for different sources. In this case, node i weights the information that it collects as intermediate node more, as defined by (5.2) and (5.1). The larger the T-Distance value, the more node i comes to believe that node m exhibits Byzantine behavior, since it behaves differently when it forwards data packets for different sources. In this case, node i weights the information that it collects as intermediate node less, as defined by (5.2) and (5.1). If node i has computed a value for $\widehat{T}_{i,m}$, but has not yet computed a value for $\widetilde{T}_{i,m}$, we assume that $e = 0$ and, consequently, $w = 1$.

First-hand Trust Evaluation Scheme

A node i calculates first-hand trust with respect to another node m using counters A_m and M_m. The counters A_m represent the estimated overall number of packets forwarded *correctly* (not dropped or misrouted) from node m over an observation window and thus are given as functions of the counters $\widetilde{A}_m$ and $\widehat{A}_m$. Here, the $\widetilde{A}$-counters indicate the number of packets forwarded *correctly* when the node collecting the empirical evidence is the source of the traffic flow, whereas the $\widehat{A}$-counters indicate the number of packets forwarded *correctly* when the node collecting the empirical evidence is an intermediate node on the path that a traffic flow traverses.

Similarly, counters M_m represent the total overall number of packets sent to node m for forwarding over an observation window and thus are given as a functions of the counters $\widetilde{M}_m$ and $\widehat{M}_m$, where the $\widetilde{M}$-counters indicate the total number of packets sent for forwarding when the node collecting the empirical evidence is the source of the traffic flow and the $\widehat{M}$-counters indicate the total number of packets sent for forwarding when the node collecting the empirical evidence is an intermediate node on the path that a traffic flow traverses. Then, the counters A_m and M_m are defined, respectively, as follows:

$$A_m \triangleq \sum_u \widetilde{A}_{m,u} + \sum_v w_v \widehat{A}_{m,v} - 1, \tag{5.3}$$

and

$$M_m \triangleq \sum_u \widetilde{M}_{m,u} + \sum_v w_v \widehat{M}_{m,v} - 1, \qquad (5.4)$$

where u ranges over the set of flows for which node i is the source node and v ranges over the set of flows for which node i is an intermediate node[2]. Here, the weights w_v are chosen in the range $[0,1]$, as discussed in Section 5.2.2. Finally, the trust $t_{i,m}$ and confidence $c_{i,m}$ that node i attributes to node m over an observation window are computing using (4.1) and (4.2), respectively, as follows:

$$t_{i,m} = t(A_m, M_m)$$

and

$$c_{i,m} = c(A_m, M_m),$$

from which the trustworthiness value $T_{i,m}$ is computed via (4.3).

5.2.3 Recommendations

The recommendation scheme introduced as part of the E-Hermes scheme is adopted in the BR-Hermes scheme. The proposed scheme exploits information sharing among nodes to accelerate the convergence of trust establishment procedures, yet is robust against the propagation of false trust information by malicious nodes. In Section 5.5, we compare the convergence of BR-Hermes with and without the use of recommendations under different scenarios.

5.3 Punishment Scheme

The ability of the BR-Hermes scheme to identify Byzantine behavior enables us to develop a punishment scheme for nodes that do not forward correctly the packets that they receive for forwarding. Such a scheme is needed to prevent selfish nodes from dropping packets to ensure that they are not used as intermediate nodes, hence saving their battery power.

Assume that network node i computes at a given time instance $t = k$ opinion $P_{i,m}$ for node m (see (4.13)). Let $B_f^{i,m}$, where $0 \leq B_f^{i,m} \leq 1$, denote the probability that the node i performs incorrect packet forwarding for node m in steady-state. We can extend the relationship between two nodes to define a punishment scheme for nodes. More precisely, we define the behavioral policy for node i as follows:

$$B_f^{i,m} = \begin{cases} 0, & P_{i,m} = T_{def}, \\ 1 - P_{i,m}, & \text{otherwise.} \end{cases} \qquad (5.5)$$

Under BR-Hermes, node i chooses its forwarding behavior based on its observations and computation of its opinion $P_{i,m}$ for another network node m. At initialization time, node i does not have an opinion for node m, and until it forms its opinion, node i is required to forward correctly the packets it receives from node m for forwarding. When node i forms its opinion $P_{i,m}$ for node m, node i will base its forwarding behavior on this opinion as given by (5.5). We note that a Byzantine node may not respect this policy and may base their forwarding behavior on an arbitrary set of rules.

---

[2]The subtraction of 1 on the right-hand sides of (5.3) and (5.4) ensures that the counters A_m and M_m have the correct values at system initialization time.

The behavioral policy defined by (5.5) can be viewed as a punishment scheme, since a node's packet forwarding behavior would be expected to be characterized by $B_f^{i,m} = 0$ at any time. However, it is important to punish nodes that attempt to thwart the trust establishment scheme by dropping or misrouting packets in order to ensure that they will not used as intermediate nodes in paths. Therefore, in BR-Hermes, node i is expected to forward packets *incorrectly* (drop or misroute) according to (5.5) to mitigate such malicious behavior. We note that for a trust establishment scheme to be effective, it must also be capable of adapting to the dynamic changes in the node behavior and the network topology. This issue is addressed in the Hermes scheme via the use of windowing mechanisms.

We do not define our behavioral policy by $B_f^{i,m} > 1 - P_{i,m}$ (when $P_{i,m} \neq T_{def}$), in which case node i would forward *incorrectly* for node m more packets than node m forwards *incorrectly* for node i. This would result in a vicious circle, where one node would forward more and more packets *incorrectly* for the other node in order to satisfy the behavioral policy and eventually, both nodes would not forward correctly any packets for each other, causing a denial-of-service condition. Moreover, we do not define our behavioral policy as $B_f^{i,m} < 1 - P_{i,m}$ (when $P_{i,m} \neq T_{def}$), where node i would forward *incorrectly* for node m less packets than node m forwards *incorrectly* for node i. Such a measure would not discourage node m from forwarding *incorrectly* packets for node i, since node i would continue forwarding more packets for node m than node m would forward for node i.

Consider the scenario shown in Fig. 5.2, which is similar to the scenario of Fig. 5.1. Here, node j is an intermediate node on the path R_1 and the source node of the traffic flow that traverses path R_2. Node B is an intermediate node on both paths and exhibits Byzantine behavior. In particular, node B forwards *correctly* all the packets that node i sends to it on path R_1, whereas it forwards *correctly* only 20% of the packets that node j sends to it on path R_2. The other nodes forward all the packets *correctly*.

Under the BR-Hermes scheme, the intermediate nodes on the paths follow the behavioral policy (5.5). Since at initialization time, all opinions equal T_{def}, all nodes should perform incorrect packet forwarding with probability zero. In the scenario of Fig. 5.2, assume that all nodes follow this policy, except for node B, which exhibits Byzantine behavior. At time t, node i forms opinion $P_{i,j}(t)$ for node j based on node's j *overall* forwarding behavior and node j forms opinion $P_{j,B}(t)$ for node B based on node's B *overall* forwarding behavior (as intermediate node on node's i flow on path R_1 and node's j flow on path R_2).

Suppose[3] that at time t',

$$R_3 = \{a_n = m, a_{n-1}, ..., a_{j+1} = B, a_0 = i, a_j = j, ..., a_1\},$$

where $n \geq 4$, denote a path from node m to node a_1, and

$$R_4 = \{a_{j+1} = B, a_0 = i, a_j = j, ..., a_1\},$$

where $j \geq 3$, denote a path from node B to node a_1. The source nodes m and B send 100 packets for forwarding to their respective destinations. Let us focus on node j, which performs incorrect packet forwarding for node B in steady-state with probability $B_f^{j,B}$, where $0 \leq B_f^{j,B} \leq 1$, as defined by the behavioral policy (5.5). Since at time t, node j formed

---

[3] Note that in general, the nodes are mobile.

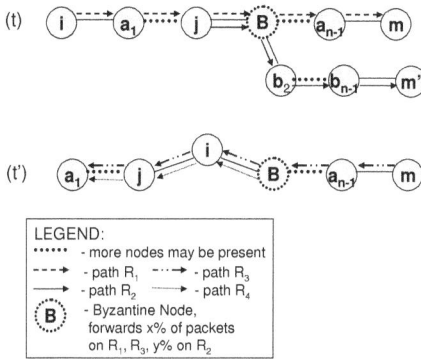

Figure 5.2: Illustration of punishment scheme in BR-Hermes.

opinion $P_{j,B}(t) \neq T_{def}$, at time t'

$$B_f^{j,B} = 1 - P_{j,B}(t).$$

Now, the key point is the opinion value that node i computes with respect to node j, after node i observes that node j performs incorrect packet forwarding for node B in steady state, with probability $B_f^{j,B}$. According to the BR-Hermes scheme, node i forms opinion $P_{i,j}(t')$ based on the information that node i collected for node j on paths $R_1, R_3,$ and R_4, since $P_{i,j}$ is a function of the counters A_j and M_j. The information that node i collected for node j on path R_1 is weighted more than the information that node i collected on paths R_3, and R_4 according to (5.3) and (5.4), since node i was the source of the flow on path R_1, but an intermediate node on paths R_3, and R_4. Therefore, even though node i observes that node j "punishes" node B by performing incorrect packet forwarding for node B in steady state with probability $B_f^{j,B}$ on path R_4, node i can recognize that the probability with which node j performs incorrect packet forwarding for it is $B_f^{j,i} \neq B_f^{j,B}$. This allows BR-Hermes to adopt the proposed behavioral policy. When the proposed behavioral policy is implemented by node x for node y, node x's behavior will not be misunderstood by other nodes (e.g., node z) that observe this behavior. The flow on path R_3 is considered to emphasize the fact that nodes are not assumed to have uniform packet forwarding behavior, which could be an unrealistic assumption. In this example, node j following the behavioral policy (5.5), performs incorrect packet forwarding for node m in steady-state with probability $B_f^{j,m} \neq B_f^{j,B}$ (where $0 \leq B_f^{j,m}, B_f^{j,B} \leq 1$).

5.4 Security Evaluation of BR-Hermes

5.4.1 Probabilistic Attacker Model

The attack space covered by the BR-Hermes scheme can be described more formally in terms of a probabilistic attacker model. The probabilistic attacker model presented in this Chapter is an extension of the probabilistic attacker model introduced in Section 3.5.3. In Chapter 3, a node was assumed to have uniform packet forwarding and trust propagating behavior towards the other network nodes. In this Chapter, a node may exhibit Byzantine behavior.

The attacker model consists of two types of attacks: 1) incorrect data packet forwarding; 2) incorrect propagation of trust information. Note that we do not distinguish among the various types of data packet forwarding misbehaviors, i.e., packet dropping, misrouting, and replay attacks. Incorrect trust propagation refers to a node which propagates a trustworthiness value that is different from the value that it should compute if it were following the BR-Hermes scheme. Thus, a node may propagate a trustworthiness value that is higher or lower than the value that a BR-Hermes-compliant node would compute.

Let $\mathcal{N}$ denote the set of all nodes in the network. A network attack scenario, in steady-state, is specified by characterizing, for each node $i \in \mathcal{N}$, the probability $B_f^{i,m}$ that the node performs incorrect packet forwarding for node $m \in \mathcal{N}$ and the probability $B_t^{i,m}$ that the node performs incorrect trust propagation for node m, where $0 \leq B_f^{i,m}, B_t^{i,m} \leq 1$. More precisely, the network attack scenario can be represented by a set of three-tuples,

$$\mathcal{S} = \{(i, B_f^{i,m}, B_t^{i,m}) : i, m \in \mathcal{N}\}. \tag{5.6}$$

Let η_f and η_t denote, respectively, thresholds on the degrees of packet forwarding and trust propagation misbehaviors that can be tolerated in the network. We set $\eta_f = \eta_t = 1 - T_{def}$.

Definition 14. Node i is **good** for node m if $B_f^{i,m} < \eta_f$.

Definition 15. Node i is **bad** for node m if $B_f^{i,m} > \eta_f$.

Definition 16. Node i is a **good recommender** for node m if $B_t^{i,m} < \eta_t$.

Definition 17. Node i is a **bad recommender** for node m if $B_t^{i,m} > \eta_t$.

A useful measure of the performance of the proposed trust establishment scheme was introduced in Section 4.6 for the E-Hermes scheme and is repeated here for convenience. The **bad node recognition** percentage or **BN-recognition** is the percentage of all bad nodes that are considered bad by all of the nodes in the network.

The BR-Hermes scheme aims to identify the set of probabilities $\{B_f^{i,m}\}$ to a sufficient degree of accuracy to distinguish between good and bad nodes, based on both first-hand information from direct observations of packet forwarding behavior and second-hand information from other nodes. The probabilistic attacker model does not preclude the possibility that nodes may collude with one another. However, the BR-Hermes scheme does not seek to identify collusions per se. Rather, the BR-Hermes scheme is able to characterize the *effect* of a colluding attack as represented by an attack scenario (5.6).

5.4.2 Security Properties of BR-Hermes

The attacker model presented above is simple, but sufficient to characterize the main security properties of the BR-Hermes scheme. Under BR-Hermes, the opinion metrics $P_{i,m}$ should closely approximate the underlying attack scenario under steady-state conditions. That is, in steady-state we should have

$$P_{i,m} \approx 1 - B_f^{m,i}, \quad \text{for all } i, m \in \mathcal{N}. \tag{5.7}$$

For the BR-Hermes framework to correctly distinguish the good nodes from the bad nodes, it is sufficient that

$$P_{i,m} > T_{def} \quad \text{for all } i \in \mathcal{N} \tag{5.8}$$

hold in steady-state. The simulation results presented in Section 5.5.2 provide validation of the steady-state properties (5.7) and (5.8).

The next few definitions were also introduced in Chapter 4 and are repeated here for convenience in the ensuing discussion on the security of the BR-Hermes scheme. A node m is **considered good** by node i when the opinion $P_{i,m} > T_{def}$. A node m is **considered bad** by node i when the opinion $P_{i,m} < T_{def}$. A node m is **considered a good recommender** by node i when the recommender trustworthiness $T_{i,m}^R > T_{def}$. A node j is **considered a bad recommender** by node i when the recommender trustworthiness $T_{i,j}^R < T_{def}$.

Under the probabilistic attacker model, the BR-Hermes scheme is able to distinguish the good nodes from the bad nodes in a network scenario with high accuracy, as demonstrated through the simulation results presented in Section 5.5.2. Nonetheless, BR-Hermes aims to calculate the opinion metrics $P_{i,m}$ accurately to closely approximate the underlying attack scenario under steady-state conditions, as characterized by (5.7). The accuracy of the Hermes family of schemes is also discussed in Section 3.7.2. Note that the probabilistic attacker model only characterizes steady-state behavior. To accommodate dynamic changes in the network attack in practice, the proper use of windowing as discussed in section 3.4.4 is necessary to maintain the responsiveness of the BR-Hermes scheme.

The key security properties provided by the BR-Hermes scheme, beyond what is provided in the previous Hermes schemes, are summarized as follows:

1. **Byzantine detection.** The proposed scheme can identify the bad nodes and bad recommenders, even when the nodes exhibit Byzantine packet forwarding and trust propagating behavior. This ability is the result of the key novel component of our scheme to weight the first-hand information that a node obtains from a traffic flow, depending on whether the given node is the source of the flow or an intermediate node on the route that the flow traverses.

2. **Byzantine robustness.** The proposed scheme is robust to the presence of bad nodes and bad recommenders, even when the nodes exhibit Byzantine packet forwarding and trust propagating behavior. Our simulation studies show very few false positives (i.e., a good node is identified as bad) and false negatives (i.e., a bad node is identified as good) even when the proportion of bad recommenders is as high as 90%. Similarly, the scheme performs well even when the proportion of bad nodes is high.

3. **Discouragement of selfish node behavior.** The proposed scheme implements a behavioral policy that discourages nodes from dropping packets to avoid being used

as intermediate nodes in paths. The accuracy of the scheme is not influenced by the implementation of the behavioral policy, as demonatrated by our simulation results (see Section 5.5).

The BR-Hermes scheme also provides the security properties provided by the previous Hermes schemes as listed below (see Chapter 4).

1. **Ability to capture independence between packet forwarding and trust propagation misbehaviors.**

2. **Resilience to the presence of bad nodes and bad recommenders.**

3. **Resilience to attacker placement.**

4. **Resilience to multiple, concurrent, and colluding attacks.**

5. **Resilience to attack frequency.**

6. **Resilience against packet duplication and replay attacks.**

5.4.3 Security Analysis

We analyze the resistance of BR-Hermes to: 1) incorrect data packet forwarding and 2) incorrect propagation of trust information attacks by Byzantine nodes. The resistance of BR-Hermes to incorrect data packet forwarding and incorrect propagation of trust information attacks by non-Byzantine nodes is the same as the resistance of E-Hermes to these attacks and is discussed in Section 4.6. Byzantine nodes perform grey-hole attacks, i.e., the Byzantine nodes drop packets of some flows that traverse them. Non-Byzantine nodes perform black hole attacks, i.e., the non-Byzantine nodes drop packets of all flows that traverse them.

Attacker nodes may impact the data flows, which traverse through them. The number of attacker nodes in the network has impact on the number of flows that may be attacked. The more the attacker nodes in the network, the more flows may be attacked. One or more attacker nodes may participate in a flow. These attackers may act independently or may collude with one another. However, the BR-Hermes scheme does not seek to identify collusions per se. Rather, the BR-Hermes scheme is able to characterize the *effect* of a colluding attack.

We shall assume that every node, whether good or bad, forwards ACK or NACK packets corresponding to packets that it has forwarded earlier. This assumption simplifies the security evaluation given below, but does not represent any limitation in the BR-Hermes scheme itself. In the BR-Hermes framework, a given node X has nothing to gain by failing to forward an ACK or NACK packet associated with a packet that it has forwarded previously. If node X fails to forward a ACK/NACK packet, node X will be penalized by all of the upstream nodes on the associated route as though it had not forwarded the original packet.

Byzantine Nodes

Fig. 5.3 illustrates the response of BR-Hermes to packet forwarding misbehavior from a single Byzantine node, labeled B, on routes $R_1 = \{Y_3, Y_2, Y_1, o, B, Z_2, \cdots\}$ and $R_2 = \{Y_2, Y_1, o, B, Z_2, \cdots\}$ corresponding to flow f_1 and f_2 respectively. In particular, node B drops packets from source

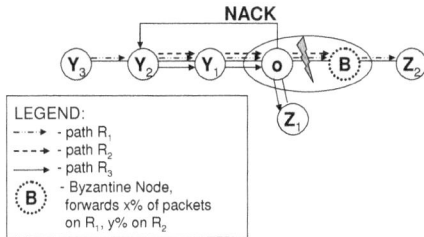

Figure 5.3: Single Byzantine node.

node Y_2 (on route R_2), but not from source node Y_3 (on route R_1). Node B's upstream neighbor, node o, observes node's B Byzantine behavior at the MAC layer (according to (5.3, 5.4)) and initiates NACKs for all packets that are not acknowledged by node B. As discussed in section 4.3, first-hand trust evaluation depends on the first-hand information gathered from a neighbor or a non-neighbor node.

- **First-hand information from neighbor node.** Node o is the FIN of Y_1 with respect to routes R_1 and R_2. Thus, node Y_1 is able to verify the correct forwarding behavior of node o, upon receiving a NACK from node o, and node Y_1 penalizes node B for the dropped packets on route R_2.

- **First-hand information from non-neighbor node.** Upon receiving a NACK initiated by node o, node Y_2 penalizes both nodes o and B. However, as a more observation data involving the two nodes with respect to different flows is accumulated over time, eventually node o will be recognized as a *good* node, whereas node B will be recognized as *bad*. For example, suppose that node Y_2 also establishes route $R_3 = \{Y_2, Y_1, o, Z_1, \cdots\}$ for its flow f_3. Node Y_2 accumulates evidence from this flow indicating that node o is a good node (for node Y_2) and identifies only node B as a bad node (for node Y_2). At the same time, nodes Y_2, Y_3 observe that node B does not drop packets of flow f_1 for source node Y_3. Nodes Y_2, Y_1, o that are upstream of node B when it exhibits Byzantine behavior, can identify node B's Byzantine behavior.

Multiple attacker nodes along a route may act independently or form collusions. Figure 5.4 illustrates the situation of packet forwarding misbehavior from multiple non-neighbor Byzantine nodes B_1 and B_2 on routes $R_1 = \{Y_3, Y_2, Y_1, o_1, B_1, o_2, B_2, Z_3, \cdots\}$ and $R_2 = \{Y_2, Y_1, o_1, B_1, o_2, B_2, Z_3, \cdots\}$ corresponding to flow f_1 and flow f_2 respectively. Nodes B_1 and B_2 drop a certain percentage of packets forwarded to them from source node Y_2 (along route R_2), but not from source node Y_3 (on route R_1). Here, the response of BR-Hermes is similar as in the case of a single Byzantine node depicted in Fig. 5.3. Note that if nodes B_1 and B_2 dropped a certain percentage of packets forwarded to them by different sources (e.g., node B_1 dropped a certain percentage of packets forwarded to it by source Y_3, whereas node B_2 dropped a certain percentage of packets forwarded to it by source Y_2), the attack scenario would be analyzed as two single Byzantine nodes on two different routes.

90

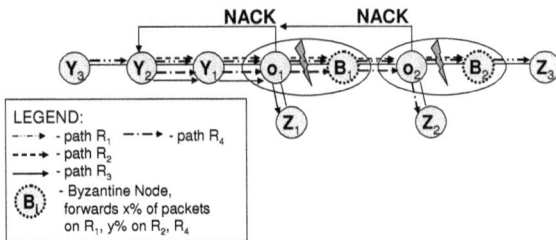

Figure 5.4: Multiple non-neighbor Byzantine nodes.

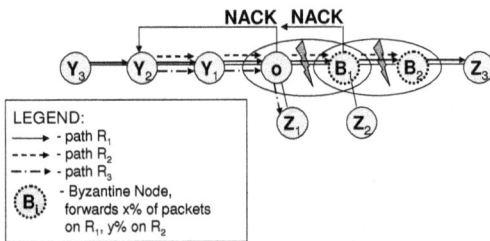

Figure 5.5: Multiple neighbor Byzantine nodes.

Node B_1's upstream neighbor node o_1 observes node B_1's Byzantine behavior at the MAC layer (according to (5.3), (5.4)) and initiate NACKs for all packets that are not acknowledged by node B_1. Similarly, node B_2's upstream neighbor node o_2, observe node B_2's Byzantine behavior at the MAC layer (according to (5.3), (5.4)) and initiate NACKs for all packets that are not acknowledged by node B_2. Node Y_1 watches on the MAC layer the forwarding behavior of node o_1 and observes that node o_1 forwards all the packets sent to it for forwarding. Thus, upon the NACK receipt from node o_1, node Y_1 knows that node B_1 is a bad node. However, node Y_1, as also node Y_2, upon a NACK receipt, which was initiated by node o_2, penalizes both nodes o_2 and B_2. Similarly, node Y_2, upon a NACK receipt, which was initiated by node o_1, penalizes both nodes o_1 and B_1. As discussed above, nodes o_1, o_2 can be recognized as good nodes only through other flows in which nodes o_1, o_2 are not penalized.

Figure 5.4 summarizes the BR-Hermes response to an incorrect data packet forwarding attack of multiple neighbor Byzantine nodes on a route. This attack scenario is similar to the previous one, with the difference being that Byzantine nodes B_1 and B_2 are neighbors on routes $R_1 = \{Y_3, Y_2, Y_1, o, B_1, B_2, Z_3, \cdots\}$ and $R_2 = \{Y_2, Y_1, o, B_1, B_2, Z_3, \cdots\}$ corresponding to flow f_1 and flow f_2 respectively. Nodes B_1 and B_2 drop a certain percentage of packets forwarded to them from source node Y_2 (along route R_2), but not from source node Y_3 (on

91

route R_1). Note that if nodes B_1 and B_2 dropped a certain percentage of packets forwarded to them by different sources, the attack scenario would be analyzed as two single Byzantine nodes on two different routes.

Node B_1's upstream neighbor node o and B_2's upstream neighbor node B_1, following the acknowledgement scheme, will initialize NACKs for all packets that are not acknowledged by nodes B_1 and B_2, respectively. Node Y_2, upon receiving a NACK initiated by node o, penalizes both nodes o and B_1. Upon receiving a NACK initiated by node B_1, node Y_2 penalizes both nodes B_1 and B_2. As discussed above, node o can be recognized as a good node only through other flows in which node o is not penalized.

Byzantine Recommender Nodes

The BR-Hermes scheme relies on the exchange of trustworthiness information among nodes through recommendations. Thus, an obvious attack on the BR-Hermes scheme would be for nodes to propagate false trustworthiness information. Incorrect trust propagation refers to a node which propagates an trustworthiness value that is different from the value that it should compute if it were following the BR-Hermes scheme. Thus, a node may propagate an trustworthiness value that is higher or lower than the value that a BR-Hermes-compliant node would compute. A Byzantine recommender sends different recommendations to different nodes. A node considers a recommender bad or good recommender based on the number of correct or incorrect recommendations sent to it by the recommender (see Section 5.4.2).

The RC-test (see Section 4.4.1) ensures that recommendations are accepted only when the recommended trustworthiness value is close to the first-hand trustworthiness value. If the first-hand trustworthiness value is computed from confidence smaller than c_{acc}, the node only temporarily accepts the maximum value from among all the recommenders. Because of this, bad recommender nodes are identified correctly, as also verified by our performance analysis in Section 5.5. Additionally, the false categorization of a recommender node as a bad recommender does not influence the correct evaluation of the nodes as good or bad. A bad recommender false positive only results in discarding the recommendations received by the recommender node, as discussed in more detail in Section 4.6.

Collusion of Byzantine Node and Bad Recommender

The collusion of a Byzantine node and bad recommender can be viewed as an attack on the flow, where the Byzantine node is a bad node and colludes with a bad recommender. This attack is analyzed in Section 4.6.

5.5 Performance Evaluation of BR-Hermes

In this section we evaluate the performance of BR-Hermes. We first discuss the communication and computational overhead of the scheme. Then, we evaluate the accuracy of BR-Hermes by presenting some representative results from our simulation experiments.

5.5.1 Communication and Computational Overhead

The extensions discussed in this Chapter, which enable BR-Hermes to detect Byzantine behavior and implement the proposed behavioral policy, do not impose any additional communication overhead beyond that of E-Hermes, which is discussed in Chapter 4. The computational overhead of E-Hermes was also presented in Chapter 4. The extensions discussed in this Chapter, which enable BR-Hermes to detect Byzantine behavior and implement the proposed behavioral policy, impose little additional computational overhead. The acknowledgement-related and recommendation-related computational overhead of E-Hermes is not increased by the BR-Hermes extensions.

In BR-Hermes, each node maintains a set of values associated with each of the other nodes in the network:

- Counters $\widehat{A}, \widetilde{A}, A$ and $\widehat{M}, \widetilde{M}, M$;

- Trust value $\widehat{t}, \widetilde{t}, t$ and confidence value $\widehat{c}, \widetilde{c}, c$;

- Trustworthiness value $\widehat{T}, \widetilde{T}, T$;

- Opinion value P;

- Recommender counters A^R and M^R;

- Recommender trustworthiness value T^R.

When a given node x is the source of a flow, node x updates a subset of the counters $\widehat{A}$ and $\widehat{M}$ each time it sends a packet to node y for forwarding. When node x is an intermediate node on the path that a flow traverses, node x updates a subset of counters $\widetilde{A}$ and $\widetilde{M}$ each time it sends a packet to node y for forwarding. For such a given packet, the counters $\widehat{A}, \widetilde{A}$ and $\widehat{M}, \widetilde{M}$ associated with the downstream nodes on the route are updated in accordance with the scheme for gathering first-hand trust information discussed in Section 4.3. The values of counters A, and M, trust $\widetilde{t}, \widehat{t}$, and t, confidence $\widetilde{c}, \widehat{c}$, and c, trustworthiness $\widetilde{T}, \widehat{T}$, and T and opinion P are then updated whenever the $\widehat{A}, \widetilde{A}$ and $\widehat{M}, \widetilde{M}$ counters are updated. The values A^R, M^R, and T^R are updated whenever the RC-test is applied (see Section 4.4.1). To be responsive to changes in the network dynamics, a windowing mechanism should be applied in computing the trustworthiness and opinion values (see Section 3.4.4). For example, if a window of size K is used in computing the opinion value P, then the K most recent values of P over the averaging window must be stored. In summary, the storage and the computational requirements for maintaining the trustworthiness and opinion values in BR-Hermes are relatively modest.

5.5.2 Performance Results

Simulation methodology

We present some representative results from our simulation experiments for evaluating the accuracy of our scheme under different network and attack scenarios. The network consists of 50 nodes that are randomly placed in a 3000 m by 3000 m area. The wireless radio transmission range of the nodes is set to 250 m. The presented results suggest the effectiveness of BR-Hermes in large network scenarios. Nodes exhibit four types of behavior.

- Type I: Good nodes and good recommenders;

- Type II: Bad nodes and good recommenders;

- Type III: Good nodes and bad recommenders;

- Type IV: Bad nodes and bad recommenders.

Randomly chosen nodes are set to exhibit Byzantine behavior, i.e., they exhibit a different type of behavior towards different nodes.

A predefined number of flows is generated for each simulation scenario. The route corresponding to a flow is not derived based on a given topology, but is chosen randomly to reflect the network topology at a given point in time. Thus, the effect of a dynamically changing network topology is captured in the simulation. The nodes in the network collect empirical evidence and build their trustworthiness and opinion values for all other network nodes based on traffic generated by the traffic flows.

Since the traffic flows are generated randomly, one or more misbehaving nodes may participate per flow. Misbehaving nodes may be neighbors or non-neighbors. We remark that in the simulations discussed here, we do not employ the averaging windows introduced in Section 3.4.4, in order to simplify the presentation of results. Implementation of the averaging windows would have further improved the accuracy of the final opinions when the node behaviors change over time (see Fig. 5.8). The presented results suggest the effectiveness of BR-Hermes in relatively large network scenarios of 50 nodes.

Network View without Behavioral Policy

In the first simulation scenario, the BR-Hermes scheme (with recommendations) runs without the behavioral policy being implemented. 800 flows are established along different paths in the network. The minimum and maximum number of nodes allowed on a route are five and seven respectively. Nodes $21 - 50$ are assigned to be of Type I for all nodes. They forward 100% of the packets that they should be forwarding and propagate correct opinion P. Nodes $1 - 5$ are assigned to be of Type II for all nodes and nodes $6 - 10$ are assigned to be of Type II for nodes $1 - 25$, whereas they are of Type I for nodes $26 - 50$. Thus, nodes $6 - 10$ exhibit Byzantine behavior. Nodes of Type II forward 20% of the packets received for forwarding, but propagate correct opinion P.

Nodes $16 - 20$ are assigned to be of Type III. Nodes of Type III forward 100% of the packets received for forwarding, but propagates recommendations of fixed opinion $P = 0.5$. Nodes $11 - 15$ are chosen to be of Type IV for nodes $1 - 25$, whereas they are of Type III for nodes $26 - 50$. Thus, nodes $11 - 15$ exhibit Byzantine behavior. Nodes of Type IV forward 20% of the packets received for forwarding, and propagate recommendations of fixed opinion $P = 0.5$. Although, in this case 40% of the nodes exhibit malicious behavior of one or another type, increasing this percentage does not affect the ability of the BR-Hermes scheme to form accurate opinions. The source nodes send 100 data packets during each observation window W (also called "round"). The trustworthiness parameter r is set as $r = \sqrt{2/9}$, and the RC-test threshold η (see Section 4.4.1) is set to 0.1. Recommendations are exchanged among nodes that are in the same route and between any two nodes given that one of the nodes has formed opinions for nodes that the other node wants to use as intermediate nodes on a route.

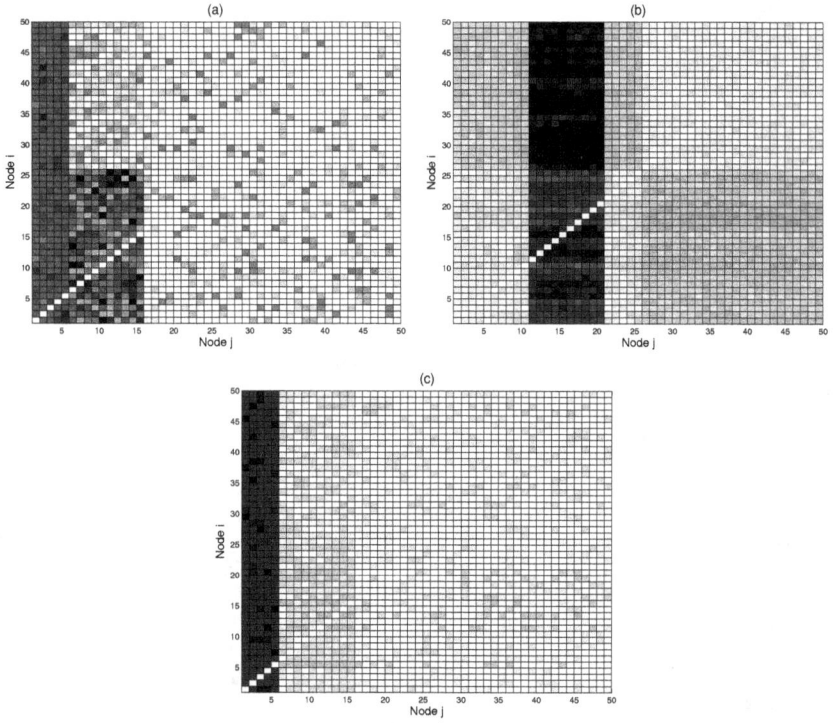

Figure 5.6: Network view: (a) Opinion $P_{i,j}$ for BR-Hermes, (b) Recommender trustworthiness $T_{i,j}^R$ for BR-Hermes, (c) Opinion $P_{i,j}$ for E-Hermes.

Fig. 5.6 illustrates the opinion value that node i places on node j with a gray-scale representation. A black color implies an opinion value of 0, white represents an opinion value of 1, while intermediate values are represented by different shades of gray. Fig. 5.6 (a) illustrates the opinion values, $P_{i,j}$ which is the opinion formed in terms of packet forwarding. One can see that nodes $1 - 5$ are correctly identified as bad nodes by all other nodes (no false negatives). Nodes $16 - 50$ are also correctly identified as good nodes by all other nodes (no false positives). Nodes $6 - 15$ are correctly identified as good by nodes $26 - 50$ (no false positives), whereas nodes $6 - 15$ are in their majority correctly identified as bad nodes by nodes $1 - 25$ (53 false negatives). In total, 2450 opinions have been formed (50 nodes formed opinions about all other 50 nodes) and there were 53 false positives and negatives, which equals to 2.163% of false positives and negatives.

Fig. 5.6 (b) shows the recommender trustworthiness values, $T_{i,j}^R$, which are the opinions formed in terms of trust propagation. Nodes $11 - 20$ are correctly identified as bad recommenders by all other nodes. The remaining nodes are correctly identified as good recommenders. Note that nodes $26 - 50$ consider nodes $26 - 50$ as slightly better recommenders than nodes $1 - 25$. This is because nodes $26 - 50$ agree in their opinions about all other network nodes (nodes $1 - 50$), but do not agree with nodes $1 - 25$ in their opinions about all other network nodes (in particular, nodes $26 - 50$ do not agree with nodes $1 - 25$ in their opinions about nodes $6 - 15$). Similarly nodes $1 - 25$ consider nodes $1 - 25$ as slightly better recommenders than nodes $26 - 50$. This is because nodes $1 - 25$ agree in their opinions about all other network nodes (nodes $1 - 50$), but do not agree with nodes $26 - 50$ in their opinions about all other network nodes.

Fig. 5.6 (c) illustrates the opinion values, $P_{i,j}$ when E-Hermes, which does not exhibit Byzantine detection is implemented for the same simulation scenario. The bad nodes are recognized correctly, but not the Byzantine nodes. This is (as explained in details in section 5.2.1) because nodes do not differentiate the evidence they accumulate when they are source nodes of the flows from the evidence they accumulate when they are intermediate nodes of flows. As a result, the Byzantine nodes are viewed as good nodes always. Comparing (a) and (c), we see that BR-Hermes outperforms E-Hermes. We have also tested our scheme under various attack scenarios, varying the number of bad recommenders and bad nodes, and found that BR-Hermes forms accurate opinions in all cases.

Network View of BR-Hermes

In the next simulation scenario, the BR-Hermes scheme (with recommendations) runs while the behavioral policy is implemented. The simulation scenario of Section 5.5.2 is implemented.

Fig. 5.7 (a) illustrates the opinion values $P_{i,j}$ with a gray-scale representation. One can see that nodes $1-5$ are correctly identified as bad nodes by all nodes (no false negatives) and that symmetrically nodes $1-5$ consider all nodes bad. This is the result of the implemented behavioral policy (5.5). All nodes that consider nodes $1-5$ bad, behave to nodes $1-5$ as nodes $1-5$ behave towards them, and therefore nodes $1-5$ consider them bad.

Nodes $6-15$ are correctly identified as good by nodes $26-50$ (no false positives), whereas nodes $6-15$ are in their majority correctly identified as bad nodes by nodes $1-25$ (38 false negatives). Symmetrically, nodes $6-15$ consider the majority of nodes $1-25$ bad (16 false negatives), which is the result of the implemented behavioral policy (5.5). All nodes that consider nodes $6-15$ bad, behave to nodes $6-15$ as nodes $6-15$ behave towards them, and

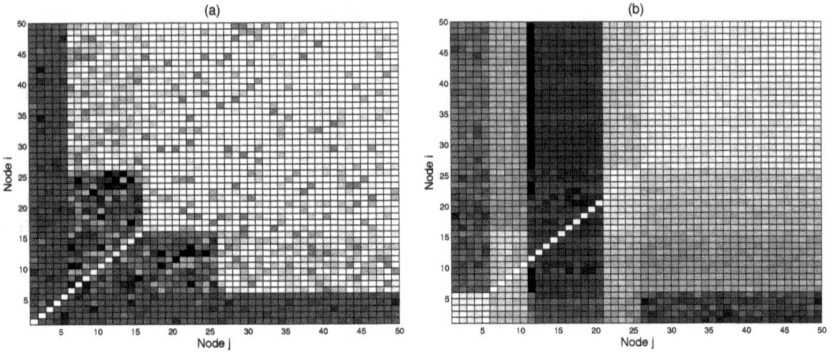

Figure 5.7: Network view of BR-Hermes: (a) Opinion $P_{i,j}$, (b) Recommender trustworthiness $T_{i,j}^R$.

therefore nodes $6 - 15$ consider them bad.

Nodes $16 - 50$ are also correctly identified as good nodes by nodes $16 - 50$ and nodes $6 - 15$ (no false positives). In total, 2450 opinions have been formed (50 nodes formed opinions about all other 50 nodes) and there were 54 false positives and negatives, which equals to 2.2% of false positives and negatives. Note that the behavioral policy does not cause false positives, due to BR-Hermes' ability to detect Byzantine behavior.

Fig. 5.6 (b) shows the recommender trustworthiness values, $T_{i,j}^R$, which are the opinions formed in terms of trust propagation. Nodes $11 - 20$ are correctly identified as bad recommenders by all other nodes. Nodes $26 - 50$ consider nodes $26 - 50$ as good recommenders and better recommenders than nodes $1 - 25$. In particular, nodes $26 - 50$ consider nodes $26 - 50$ as the best recommenders, nodes $21 - 25$ as good recommenders but slightly worse recommenders than nodes $26 - 50$, nodes $11 - 20$ bad recommenders, nodes $6 - 10$ as good recommenders but slightly worse recommenders than nodes $21 - 25$, and nodes $1 - 5$ worse recommenders than nodes $6 - 10$, better recommenders than nodes $11 - 20$, but also as bad recommenders. This is because nodes $26 - 50$ agree in their opinions about all other network nodes (nodes $1 - 50$), but do not agree with nodes $1 - 25$ in their opinions about all other network nodes. In particular, nodes $26 - 50$ do not agree with nodes $21 - 25$ in their opinions about nodes $6 - 15$, with nodes $11 - 20$ in their opinions about all nodes, with nodes $6 - 10$ in their opinions about nodes $6 - 25$ and with nodes $1 - 5$ in their opinions about nodes $5 - 50$.

Nodes $16 - 25$ consider nodes $21 - 25$ as good recommenders and better recommenders than all other nodes. In particular, nodes $16 - 25$ consider nodes $21 - 25$ as the best recommenders, nodes $26 - 50$ as good recommenders but slightly worse recommenders than nodes $21 - 25$, nodes $11 - 20$ bad recommenders, nodes $6 - 10$ as good recommenders but slightly worse recommenders than nodes $26 - 50$, and nodes $1 - 5$ worse recommenders than nodes $6 - 10$, better recommenders than nodes $11 - 20$, but also as bad recommenders. This is because nodes $16 - 25$ agree in their opinions about all other network nodes (nodes $1 - 50$), but

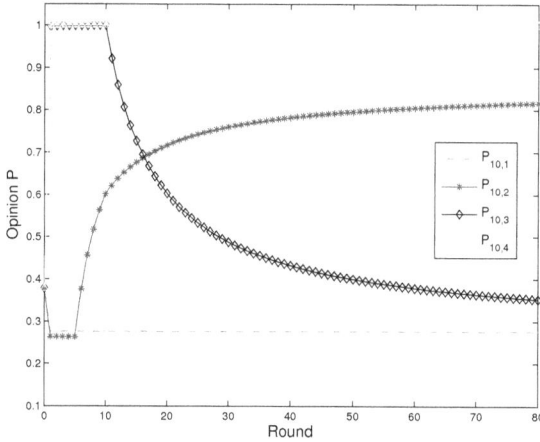

Figure 5.8: Opinion that node 10 forms for nodes $1, 2, 3, 4$ from round 1 to 80. Nodes $2, 3$ change their forwarding behavior in rounds 5 and 10 respectively.

do not always agree with the other nodes. In particular, nodes $16 - 25$ do not agree with nodes $26 - 50$ in their opinions about nodes $6 - 15$, with nodes $11 - 20$ in their opinions about all nodes, with nodes $6 - 10$ in their opinions about nodes $16 - 25$ and with nodes $1 - 5$ in their opinions about nodes $5 - 50$.

Similarly, nodes $6 - 10$ and nodes $1 - 5$ categorize the other nodes in 5 different groups in terms of their trustworthiness as recommenders, which can be explained by the relative opinions that they have about the network nodes. All nodes agree that nodes $11 - 20$ are bad recommenders.

Adaptive Behavior

To demonstrate the ability of BR-Hermes, with the punishment scheme, to adapt to changes in the node behaviors, we use the same simulation scenario. A set of 100 flows are generated and the source nodes send 100 data packets during each round. The simulation runs for eighty rounds. However, now nodes $4, 6 - 10$ are of Type I. Nodes $3, 5$ are bad recommenders, propagating opinions with value $P = 0.5$. Nodes $1, 2$ are of Type II. Node 3 is good for rounds 1-10 and then becomes bad, thus switching from Type III to Type IV. Node 2 is bad for rounds 1-5 and then becomes good, thus switching from Type II to Type I. Node 5 is of Type III. Good nodes forward 100% of the packets that they should be forwarding. Bad nodes forward 20% of the packets received for forwarding. As before, the RC-test threshold η is set to 0.1.

The opinions P that node 10 places on nodes $1, 2, 3, 4$ over 80 rounds is shown in Fig. 5.8. Our scheme accurately evaluates trust and adapts to changes in the nodes' behaviors. Note

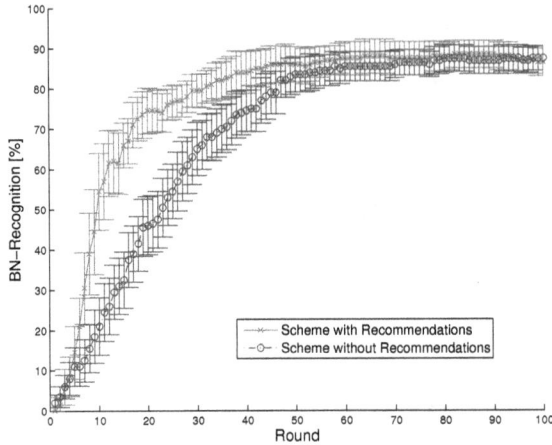

Figure 5.9: Convergence comparison of scheme with and without recommendations in respect to BN-recognition.

that the past behavior of a node influences the value of the current opinion P. For example, at round 80 $P_{10,4} \approx 1$, whereas $P_{10,2} \approx 0.82$. Similarly, at round 80 $P_{10,1} \approx 0.27$, whereas $P_{10,2} \approx 0.35$. The implementation of the windowing mechanisms would systematically expire old observation data in order to improve the responsiveness of the system. We remark that the ability of BR-Hermes scheme to quickly adapt to changing node behavior is a key feature that makes it practical for real-world networks.

Convergence Comparison

In the next simulation to be presented, we compare the convergence of our scheme with and without the use of recommendations. The objective is to investigate the BN-recognition of our scheme as a function of active network flows. The simulated network consists of 10 nodes. Node 1 is a bad node. Node 3 exhibits Byzantine behavior and is bad node for nodes $1-5$ and good for nodes $6-10$. Nodes 2 also exhibits Byzantine behavior and is good for nodes $1-5$ and bad for nodes $6-10$. The rest of the nodes $4-10$ are good nodes. Bad recommenders are nodes $3,4$ for the scheme with recommendations. As in earlier simulations, good nodes forward 100% of packets, bad nodes 20%, good recommenders propagate valid trust values, whereas bad recommenders send $P = 0.5$. Initially one flow is generated and then one flow is added per round. The flows are randomly generated. The number of nodes on a route is set to 5.

Figure 5.9 shows the BN-recognition of the scheme with and without recommendations. The error bars indicate the 90% confidence intervals obtained from executing on the order of 20 simulation trials for each estimated value. As expected, recommendations accelerate the convergence of the trust establishment procedures. For example, with recommendations, the BN-recognition exceeds 65% at 15 rounds, whereas when recommendations are not used the BN-recognition exceeds 65% at 32 rounds, whereas at 15 rounds it is 32%. The BN-recognition converges to a steady-state value of more than 87%.

5.6 BR-Hermes Summary

We presented BR-Hermes, a Byzantine robust trust establishment scheme for MANETs, which is designed to improve the reliability of packet forwarding over multi-hop routes, particularly in the presence of Byzantine nodes. In the proposed scheme, each node determines the trustworthiness of the other nodes with respect to reliable packet forwarding by combining first-hand trust information obtained independently of other nodes and second-hand trust information obtained via recommendations from other nodes. BR-Hermes exploits information sharing among nodes to accelerate the convergence of trust establishment procedures, yet is robust against the propagation of false trust information by Byzantine nodes.

The proposed BR-Hermes scheme extends the E-Hermes framework introduced in Chapter 4 in several important ways. In the BR-Hermes, first-hand information that a node obtains from a traffic flow is weighted depending on whether the given node is the source or an intermediate node on the route that the flow traverses. This extension enables BR-Hermes to detect Byzantine behavior and allows us to introduce a punishment scheme that discourages selfish node behavior.

The BR-Hermes scheme allows nodes to form accurate opinions for any network node independent of whether it exhibits Byzantine behavior. The number of nodes that propagate false trust information does not influence the robustness of the system. Three types of malicious node behavior are identified: (i) dropping or misrouting packets but propagating true opinion values, (ii) forwarding packets but propagating false opinion values, and (iii) dropping or misrouting packets and propagating false opinion values. The effect of attacks by malicious nodes is identified under BR-Hermes, either when they operate separately or form collusions. Our simulation results demonstrate the effectiveness of the BR-Hermes scheme in distinguishing among malicious and non-malicious nodes in a variety of network scenarios involving Byzantine nodes that are malicious both with respect to packet forwarding and trust propagation.

Chapter 6: Concluding Remarks

We propose Hermes, a quantitative trust establishment framework for MANETs, which is designed to improve the reliability of packet forwarding over multi-hop routes in the presence of potentially malicious nodes. Using a Bayesian framework, two metrics are defined: trust and confidence, which are computed based on the empirical first-hand observations of packet forwarding behavior by neighbor nodes. Trust characterizes the belief in the reliability of a neighbor node with respect to packet forwarding. The confidence value associated with a given trust value characterizes the statistical reliability of the trust value. Trust and confidence are mapped into a "trustworthiness" metric, which captures the impact of trust and confidence in a single value.

The concept of trustworthiness, initially defined only between neighbor nodes, is then extended to the notion of an *opinion* that a given node has for any arbitrary node. The opinion metric can be incorporated into MANET routing protocols such as DSR or AODV to improve the reliability of packet delivery in a transparent manner. A windowing scheme is used to expire old observation data in order to improve the accuracy of the opinion metric. The overhead imposed by the Hermes scheme is mainly computational. Nodes following the Hermes scheme collect statistics based on first-hand observations of packet transmissons on the wireless broadcast channel and compute the trust metrics. The communication overhead due to the propagation of second-hand trust information can be minimized by piggybacking trust information onto the routing control packets.

A probabilistic attacker model was proposed to characterize the security properties of Hermes. Our simulation experiments demonstrate the effectiveness of the Hermes framework in distinguishing among bad and good nodes as well as in the selection of more "trustworthy" routes for reliable packet delivery.

We investigated extensions to the Hermes framework to deal with the behavior of nodes that propagate invalid trustworthiness information. The extended framework is called E-Hermes. In the E-Hermes scheme, first-hand information for non-neighbor nodes is obtained via feedback from acknowledgements sent in response to data packets. The E-Hermes exploits information sharing among nodes to accelerate the convergence of trust establishment procedures. Second-hand trust information is obtained via recommendations from cooperative nodes. The trustworthiness of the recommendations and recommenders is evaluated. The concept of trustworthiness is then extended to the notion of an *opinion* that a given node has about the forwarding behavior of any arbitrary node by combining first-hand and second-hand trust information.

The proposed extensions to Hermes allow nodes to form accurate opinions for any network node and provides robustness against the propagation of false trust information by malicious nodes. The number of nodes that propagate false trust information does not influence the robustness of the system. Three types of malicious node behavior are identified: (i) dropping or misrouting packets but propagating true opinion values, (ii) forwarding packets but propagating false opinion values, and (iii) dropping or misrouting packets and propagating false opinion values. The effect of attacks by malicious nodes is identified either when they operate separately or form collusions. We presented simulation results which demonstrate the effectiveness of the E-Hermes scheme in distinguishing among malicious and non-malicious nodes in a variety of network scenarios involving nodes that are malicious both with respect to packet forwarding and trust propagation.

Additionally, we presented a further extension to the Hermes framework called BR-Hermes, a Byzantine robust trust establishment scheme for MANETs, which is designed to improve the reliability of packet forwarding over multi-hop routes, particularly in the presence of Byzantine nodes. In the proposed scheme, each node determines the trustworthiness of the other nodes with respect to reliable packet forwarding by combining first-hand trust information obtained independently of other nodes and second-hand trust information obtained via recommendations from other nodes. BR-Hermes exploits information sharing among nodes to accelerate the convergence of trust establishment procedures, yet is robust against the propagation of false trust information by Byzantine nodes.

The proposed BR-Hermes scheme extends the E-Hermes framework in several important ways. In the BR-Hermes, first-hand information that a node obtains from a traffic flow is weighted depending on whether the given node is the source or an intermediate node on the route that the flow traverses. This extension enables BR-Hermes to detect Byzantine behavior and allows us to introduce a punishment scheme that discourages selfish node behavior.

The BR-Hermes scheme allows nodes to form accurate opinions for any network node independent of whether it exhibits Byzantine behavior. The number of nodes that propagate false trust information does not influence the robustness of the system. Three types of Byzantine node behavior are identified: (i) dropping or misrouting packets but propagating true opinion values, (ii) forwarding packets but propagating false opinion values, and (iii) dropping or misrouting packets and propagating false opinion values. The effect of attacks by malicious nodes is identified under BR-Hermes, either when they operate separately or form collusions. Our simulation results demonstrate the effectiveness of the BR-Hermes scheme in distinguishing among malicious and non-malicious nodes in a variety of network scenarios involving Byzantine nodes that are malicious both with respect to packet forwarding and trust propagation.

This book does not address the implementation of Hermes to MANET routing protocols and the convergence of trust-aware routing protocols. We have developed simulations of the Hermes family of frameworks and devised simulation attach models to test and validate their performance in Matlab. We have analyzed the communication and computational overhead of the Hermes schemes and we have shown that they impose little additional overhead. Additionally, we have discussed that the routing opinion metric could be used to choose the route probabilistically, i.e., a route would be chosen with probability proportional to the routing opinion. Such a randomized routing scheme would improve the performance of the Hermes scheme, as the flows would traverse a more diverse set of nodes, providing a richer set of first-hand observation data for computing the trust metrics. Another interesting extension of the trust establishment work in this book would be to consider giving scheduling priority to flows of packets based on the opinion that is placed on the source nodes of the flows. The opinion metric can also be used as a criterion to maintain network membership.

We believe that the Hermes framework could play a critical role in an overall security architecture for MANETs. The Hermes framework presented in this book and the distributed PKI scheme proposed in [49] are complementary and form the basis for a complete framework for secure key distribution and trust establishment in terms of reliable data packet delivery for mobile ad hoc networks.

Bibliography

[1] M. Frodigh, P. Johansson, and P. Larsson, "Wireless ad hoc networking: The art of networking without a network," *Ericsson Review*, no. 4, pp. 248–263, 2000.

[2] C. Zouridaki, M. Hejmo, B. L. Mark, R. K. Thomas, and K. Gaj, "Analysis of Attacks and Defense Mechanisms for QoS Signaling Protocols in MANETs," in *Proceedings of the 4th International Workshop on Wireless Information Systems (WIS 2005)*, May 2005, pp. 61–70.

[3] M. Deutsch, "Cooperation and Trust: Some Theoretical Notes," in *Nebraska Symposium on Motivation*, M. R. Jones, Ed. Nebraska University Press, 1962.

[4] D. B. Johnson, D. A. Maltz, and Y. C. Hu, *The Dynamic Source Routing Protocol for Mobile Ad Hoc Networks*, IETF Internet Draft, April 2003.

[5] C. E. Perkins, E. M. Belding-Royer, and S. Das, *Ad Hoc On Demand Distance Vector (AODV) Routing*, IETF RFC 3561, July 2003.

[6] H. Luo, J. Kong, P. Zerfos, S. Lu, and L. Zhang, "URSA: Ubiquitous and Robust Access Control for Mobile Ad Hoc Networks," *IEEE/ACM Transactions on Networking (ToN)*, vol. 12, no. 6, pp. 1049–1063, December 2004.

[7] S. Buchegger and J.-Y. L. Boudec, "A Robust Reputation System for P2P and Mobile Ad-hoc Networks," in *Proceedings of the 2nd Workshop on Economics of Peer-to-Peer Systems*, June 2004.

[8] G. Theodorakopoulos and J. S. Baras, "Trust Evaluation in Ad-hoc Networks," in *Proceedings of the 2004 ACM workshop on Wireless Security (WiSe '04)*, 2004, pp. 1–10.

[9] L. Zhou and Z. J. Haas, "Securing Ad Hoc Networks," *IEEE Networks Special Issue on Network Security*, November 1999.

[10] P. Papadimitratos and Z. J. Haas, "Secure message transmission in mobile ad hoc networks," *Elsevier Ad Hoc Networks Journal*, vol. 1, no. 1, Jan/Feb/March 2003.

[11] S. Ghazizadeh, O. Ilghami, and E. Sirin, "Security-aware adaptive dynamic source routing protocol," in *Proceedings of the 27th IEEE Conference on Local Computer Networks*, November 2002.

[12] Y. C. Hu, D. B. Johnson, and A. Perrig, "SEAD: Secure Efficient Distance Vector Routing for Mobile Wireless Ad Hoc Networks," in *Proceedings of the 4th IEEE Workshop on Mobile Computing Systems and Applications (WMCSA '02)*, June 2002, pp. 3–13.

[13] Y. C. Hu, A. Perrig, and D. B. Johnson, "Ariadne: A secure on-demand routing protocol for ad hoc networks," in *Proceedings of the ACM MobiCom '02*. ACM SIGMOBILE, September 2002.

[14] P. Papadimitratos and Z. J. Haas, "Secure routing for mobile ad hoc networks," in *Proceedings of the SCS Communication Networks and Distributed Systems Modeling and Simulation Conference (CNDS) 2002*, January 2002.

[15] M. G. Zapata, "Secure Ad hoc On-Demand Distance Vector Routing," *ACM Mobile Computing and Communications Review (MC2R)*, vol. 6, no. 3, pp. 106–107, July 2002.

[16] S. Marsh, "Formalizing trust as a computational concept," Ph.D. dissertation, University of Stirling, 1994.

[17] A. Abdul-Rahman and S. Hailes, "Supporting trust in virtual communities," in *Proceedings of the IEEE the Hawaii International Conference on System Sciences*, January 2000.

[18] A. A. Pirzada and C. McDonald, "Establishing trust in pure ad-hoc networks," in *Proceedings of the 27th Australasian Computer Science Conference (ACSC04)*, January 2004, pp. 47–54.

[19] B. Yu and M. P. Singh, "A social mechanism of reputation management in electronic communities," in *Proceedings of the 4th International Workshop on Cooperative Information Agents*, vol. 1860. LNCS, 2000, pp. 154–165.

[20] P. Resnick, K. Kuwabara, R. Zeckhauser, and E. Friedman, "Reputation systems," *Communications of the ACM*, vol. 43, no. 12, pp. 45–48, Dec. 2000.

[21] W. Zhao, V. Varadharajan, and G. Bryan, "Type and Scope of Trust Relationships in Collaborative Interactions in Distributed Environments," in *Proceedings of the 7th International Conference on Enterprise Information Systems (ICEIS)*, May 2005.

[22] S. Marti, P. Ganesan, and H. Garcia-Molina, "Sprout: P2P routing with social networks," in *Proceedings of International Workshop on Peer-to-Peer Computing and DataBases*, 2004, pp. 425–435.

[23] S. Buchegger and J.-Y. L. Boudec, "Nodes Bearing Grudges: Towards Routing Security, Fairness, and Robustness in Mobile Ad Hoc Networks," in *Proceedings of the 10th Euromicro PDP (Parallel, Distributed and Network-based Processing)*, 2002, pp. 403 – 410.

[24] G. Theodorakopoulos and J. S. Baras, "On trust models and trust evaluation metrics for ad hoc networks," *IEEE Journal on Selected Areas in Communications*, vol. 24, no. 2, pp. 318–328, February 2006.

[25] Y. Sun, W. Yu, Z. Han, and K. J. R. Liu, "Trust modeling and evaluation in ad hoc networks," in *Proceedings of Globecom 2005*. IEEE, 2005, pp. 1862–1867.

[26] ——, "Information theoretic framework of trust modeling and evaluation for ad hoc networks," *IEEE Selected Areas in Communications*, vol. 24, no. 2, pp. 305–317, Feb. 2006.

[27] L. Eschenauer, V. D. Gligor, and J. Baras, "On trust establishment in mobile ad-hoc networks," in *Proceedings of the Security Protocols Workshop*, vol. 2845. LNCS, April 2002, pp. 47–66.

[28] T. Jiang and J. S. Baras, "Ant-based Adaptive Trust Evidence Distribution in MANET," in *Proceedings of the 2nd International Workshop on Mobile Distributed Computing (MDC)*, March 2004.

[29] J. Baras and T. Jiang, "Cooperative Games, Phase Transition on Graphs and Distributed Trust in MANET," in *Proceedings of the 43rd IEEE Conference on Decision and Control*, June 2004.

[30] D. Subramanian, P. Druschel, and J. Chen, "Ants and Reinforcement Learning: A case Study in Routing in Dynamic Networks," in *Proceedings of the 15th International Conference on Artificial Intelligence*, 1997, pp. 832–838.

[31] T. Jiang and J. S. Baras, "Autonomous Trust Establishment," in *Proceedings of the 2nd International Network Optimization Conference*, 2005.

[32] R. K. Nekkanti and C. Lee, "Trust based adaptive on demand ad hoc routing protocol," in *Proceedings of the 42nd ACM Southeast Regional Conference*, 2004, pp. 88–93.

[33] L. Buttyan and J.-P. Hubaux, "Stimulating Cooperation in Self-Organizing Mobile Ad Hoc Networks," *Mobile Networks and Applications*, vol. 8, no. 5, pp. 579–592, 2003.

[34] R. Anderson and M. G. Kuhn, "Tamper resistance - a cautionary note," in *Proceedings of the 2nd USENIX Workshop on Electronic Commerce*, November 1996, pp. 1–11.

[35] ——, "Low cost attacks on tamper resistant devices," in *Proceedings of the 5th International Workshop on Security Protocols (IWSP)*, vol. 1361. LNCS, April 1997, pp. 125–136.

[36] L. Capra, "Engineering Human Trust in Mobile System Collaborations," in *Proceedings of the 12th ACM SIGSOFT International Symposium on Foundations of Software Engineering*, 2004, pp. 107–116.

[37] M. Virendra, M. Jadliwala, M. Chandrasekaran, and S. Upadhyaya, "Quantifying Trust in Mobile Ad-Hoc Networks," in *Proceedings of the International Conference on Integration of Knowledge Intensive Multi-Agent Systems (KIMAS '05: Modeling, Evolution and Engineering)*, 2005.

[38] P. Michiardi and R. Molva, "Core: A collaborative reputation mechanism to enforce node cooperation in mobile ad hoc networks," in *Proceedings of 6th Communication and Multimedia Security Conference (CMS'2002)*, August 2002, pp. 107–121.

[39] Y. Rebahi, V. Mujica, and D. Sisalem, "A reputation-based trust mechanism for ad hoc networks," in *Proceedings of the 10th IEEE Symposium on Computers and Communications (ISCC'05)*. IEEE Computer Society, 2005, pp. 37–42.

[40] C. Ngai and M. Lyu, "Trust- and Clustering-Based Authentication Services in Mobile Ad Hoc Networks," in *Proceedings of the 2nd International Workshop on Mobile Distributed Computing (MDC'04)*, March 2004.

[41] Z. Liu, A. W. Joy, and R. A. Thompson, "A Dynamic Trust Model for Mobile Ad Hoc Networks," in *Proceedings of the 10th IEEE International Workshop on Future Trends of Distributed Computing Systems (FTDCS'04)*, 2004, pp. 80–85.

[42] S. Jin, C. Park, D. Choi, K. Chung, and H. Yoon, "Cluster-Based Trust Evaluation Scheme in an Ad Hoc Network," *ETRI Journal*, vol. 27, no. 4, pp. 465–468, August 2005.

[43] F. Kargl, S. Schlott, and M. Weber, "Identification in ad hoc networks," in *Proceedings of the 39th Annual Hawaii International Conference on System Sciences (HICSS '06)*. ACM, 2006, p. 233.3.

[44] F. Kargl, A. Geis, S. Schlott, and M. Weber, "Secure Dynamic Source Routing," in *Proceedings of the 38th Annual Hawaii International Conference on System Sciences (HICSS'05)*. ACM, January 2005, p. 320.3.

[45] F. Kargl, A. Klenk, S. Schlott, and M. Weber, "Advanced detection of selfish or malicious nodes in ad hoc networks," in *Proceedings of 1st European Workshop on Security in Ad-Hoc and Sensor Networks (ESAS)*, 2004, pp. 152–165.

[46] M. Morvan and S. Sene, "A distributed trust diffusion protocol for ad hoc networks," in *Proceedings of the International Multi-Conference on Computing in the Global Information Technology (ICCGI '06)*. IEEE Computer Society, 2006, p. 87.

[47] A. Karygiannis, E. Antonakakis, and A. Apostolopoulos, "Detecting critical nodes for manet intrusion detection systems," in *Proceedings of the 2nd International Workshop on Security, Privacy and Trust in Pervasive and Ubiquitous Computing (SECPERU '06)*. IEEE Computer Society, 2006, pp. 7–15.

[48] G. F. Marias, K. Papapanagiotou, V. Tsetsos, O. Sekkas, and P. Georgiadis, "Integrating a trust framework with a distributed certificate validation scheme for manets," *EURASIP Journal on Wireless Communications and Networking*, vol. 2006, pp. Article ID 78259, 18 pages, 2006.

[49] C. Zouridaki, B. L. Mark, K. Gaj, and R. K. Thomas, "Distributed CA-based PKI for Mobile Ad hoc Networks using Elliptic Curve Cryptography," in *Proceedings of the 1st European PKI Workshop on Research and Applications (EuroPKI 2004)*, vol. 3093. LNCS, June 2004, pp. 232–245.

[50] "Pretty good privacy," http://www.pgpi.org/.

[51] S. Capkun, L. Buttyan, and J. P. Hubaux, "Self-organized public-key management for mobile ad hoc networks," Swiss Federal Institute of Technology, Lausanne, Tech. Rep. EPFL/IC/200234, June 2002.

[52] J. P. Hubaux, L. Buttyan, and S. Capkun., "The quest for security in mobile ad hoc networks," in *Proceedings of the ACM MobiHoc*, 2001.

[53] F. Stajano and R. Anderson, "The Resurrecting Duckling: Security Issues for Ad-hoc Wireless Networks," in *Proceedings of the 7th International Workshop in Security Protocols*. LNCS, 1999, pp. 172–194.

[54] F. Stajano, "The resurrecting duckling - what next?" *Lecture Notes in Computer Science*, vol. 2133, pp. 204–214, 2001.

[55] D. Balfanz, D. K. Smetters, P. Stewart, and H. C. Wong, "Talking to strangers: Authentication in ad-hoc wireless networks," in *Proceedings of the Symposium on Network and Distributed Systems Security (NDSS '02)*, 2002.

[56] C. P. Chang, J. C. Lin, and F. Lai, "Trust-group-based authentication services for mobile ad hoc networks," in *Proceedings of the 1st International Symposium on Wireless Pervasive Computing*, 2006, pp. 4.–.

[57] Y. A. Huang and W. Lee, "A cooperative intrusion detection system for ad hoc networks," in *Proceedings of the 1st ACM Workshop on Security of Ad Hoc and Sensor Networks (SASN'03)*, Oct. 2003.

[58] C. Y. Tseng, P. Balasubramanyam, C. Ko, R. Limprasittiporn, J. Rowe, and K. Levitt, "A Specification-Based Intrusion Detection System For AODV," in *Proceedings of the 1st ACM Workshop on Security of Ad Hoc and Sensor Networks (SASN'03)*, 2003.

[59] Y. Li and J. Wei., "Guidelines on Selecting Intrusion Detection Methods in MANET," in *Proceedings of ISECON 2004*, 2004.

[60] B. Awerbuch, D. Holmer, C. Nita-Rotaru, and H. Rubens, "An on-demand secure routing protocol resilient to byzantine failures," in *Proceedings of the ACM workshop on Wireless Security (WiSE '02)*, 2002, pp. 21–30.

[61] I. Avramopoulos, H. Kobayashi, R. Wang, and A. Krishnamurthy, "Highly secure and efficient routing," in *Proceedings of the IEEE Infocom 2004*, March 2004.

[62] C. Zouridaki, B. L. Mark, M. Hejmo, and R. K. Thomas, "A Quantitative Trust Establishment Framework for Reliable Data Packet Delivery in MANETs," in *Proceedings of the 3rd ACM Workshop on Security of Ad Hoc and Sensor Networks (SASN'05)*, November 2005, pp. 1–10.

[63] ——, "Hermes: A Quantitative Trust Establishment Framework for Reliable Data Packet Delivery in MANETs," *Journal of Computer Security, Special Issue on Security of Ad Hoc and Sensor Networks*, vol. 15, no. 1, pp. 3–38, January 2007.

[64] N. Asokan and P. Ginzboorg, "Key agreement in ad-hoc networks," *Computer Communications Journal*, vol. 23, no. 17, pp. 1627–1637, 2000.

[65] S. Capkun, L. Buttyan, and J. P. Hubaux, "Self-organized public-key management for mobile ad hoc networks," *IEEE Transactions on Mobile Computing*, vol. 2, no. 1, pp. 52–64, June 2003.

[66] K. Sanzgiri, B. Dahill, B. N. Levine, C. Shields, and E. M. Belding-Royer, "A secure routing protocol for ad hoc networks," in *Proceedings of the 10th IEEE International Conference on Network Protocols (ICNP '02)*. IEEE Computer Society, 2002, pp. 78–89.

[67] M. G. Zapata and N. Asokan, "Securing Ad hoc Routing Protocols," in *Proceedings of the 2002 ACM Workshop on Wireless Security (WiSe 2002)*, September 2002, pp. 1–10.

[68] A. Papoulis, *Probability, Random Variables, and Stochastic Processes*. New York: McGraw-Hill, 1991.

[69] D. Bertsekas and R. Gallager, *Data Networks*, 2nd ed. Englewood Cliffs, New Jersey: Prentice Hall, 1992.

[70] C. Perkins, E. Belding-Royer, and S. Das, "Ad-hoc On-demand Distance Vector (AODV) Routing," *IETF RFC 3561*, July 2003.

[71] S. Marti, T. J. Giuli, K. Lai, and M. Baker, "Mitigating routing misbehavior in mobile ad hoc networks," in *Proceedings of the 6th annual international conference on Mobile computing and networking (MobiCom '00)*, 2000, pp. 255–265.

[72] D. Johnson and D. Maltz, "Dynamic source routing in ad hoc wireless networks," in *Mobile Computing*. Kluwer Academic Publishers, 1996, ch. 5, pp. 153–181.

[73] Y. C. Hu, A. Perrig, and D. Johnson, "Efficient security mechanisms for routing protocols," in *Proceedings of the Network and Distributed Systems Security*, 2003.

[74] ——, "Packet Leashes: A Defense Against Wormhole Attacks in Wireless Networks," in *Proceedings of the IEEE Infocom'03*, April 2003.

[75] B. Awerbuch, D. Holmer, R. Kleinberg, and H. Rubens, "Provably Competitive Adaptive Routing," in *Proceedings of the IEEE Infocom'05*, 2005.

[76] C. Zouridaki, B. L. Mark, M. Hejmo, and R. K. Thomas, "Robust Cooperative Trust Establishment for MANETs," in *Proceedings of the 3rd ACM Workshop on Security of Ad Hoc and Sensor Networks (SASN'06)*, October 2006, pp. 23–34.

[77] ——, "E-Hermes: A Robust Cooperative Trust Establishment Scheme for Mobile Ad hoc Networks," *Ad Hoc Networks*, Under review.

[78] I. Avramopoulos, H. Kobayashi, R. Wang, and A. Krishnamurthy, "Amendment to: Highly secure and efficient routing," February 2004.

[79] Y. C. Hu, A. Perrig, and D. Johnson, "Wormhole attacks in wireless networks," *IEEE Journal on Selected Areas in Communications*, vol. 24, no. 2, pp. 370–380, February 2006.

[80] D. Glynos, P. Kotzanikolaou, and C. Douligeris, "Preventing impersonation attacks in manet with multi-factor authentication," in *Proceedings of the 3rd International Symposium on Modeling and Optimization in Mobile, Ad Hoc, and Wireless Networks (WIOPT '05)*, 2005, pp. 59–64.

[81] C. Zouridaki, B. L. Mark, and M. Hejmo, "Byzantine Robust Trust Establishment for Mobile Ad hoc Networks," *Telecommunication Systems, Special Issue on Security, Privacy, and Trust for Beyond-3G Networks*, vol. 35, no. 3–4, pp. 189–206, August 2007.

[82] R. Perlman, "Network layer protocols with byzantine robustness," Ph.D. dissertation, Massachussets Institute of Technology, 1988.

Appendix A: Reproducibility Property

Bayes' Theorem for two continuous random variables X and Y is given by:

$$f(x|y) = \frac{f(y|x)f(x)}{\int_{-\infty}^{\infty} f(y|x)f(x)dx} \tag{1.1}$$

where $f(x|y)$ is the posterior probability density function (pdf) of X given $Y = y$, $f(y|x)$ is the likelihood function of X given $Y = y$, and $f(x)$ is the prior pdf of X.

Consider a sequence of random variables $R_1, R_2, \cdots$

Proposition 2. *Suppose the prior pdf for R_{k-1} denoted by $f_{k-1}(r)$, which follows a Beta distribution (see equation (3.3)), and a likelihood function that has the form of a binomial distribution with M_k successes over N_k trials (see equation (3.2)). Then the posterior pdf $f_k(r)$ of R_k given R_{k-1}, also follows a Beta distribution. If*

$$f_{k-1}(r) \sim Beta(a_{k-1}, b_{k-1}),$$

then given that $N_k = n_k$ and $M_k = m_k$ we have

$$f_k(r) \sim Beta(a_{k-1} + m_k, b_{k-1} + n_k - m_k).$$

Therefore, $f_k(r)$ is characterized by the parameters a_k and b_k, defined recursively as follows:

$$a_k = a_{k-1} + m_k \text{ and } b_k = b_{k-1} + n_k - m_k.$$

Proof.

$$f_k(r) = \frac{f_k(M_k = m_k|r, N_k = n_k)f_{k-1}(r)}{\int_0^1 f(M_k = m_k|r, N_k = n_k)f_{k-1}(r)dr} =$$

$$= \frac{\binom{n_k}{m_k}r^{m_k}(1-r)^{n_k-m_k}\frac{r^{a_{k-1}-1}(1-r)^{b_{k-1}-1}}{B(a_{k-1},b_{k-1})}}{\int_0^1 \binom{n_k}{m_k}r^{m_k}(1-r)^{n_k-m_k}\frac{r^{a_{k-1}-1}(1-r)^{b_{k-1}-1}}{B(a_{k-1},b_{k-1})}dr} =$$

$$= \frac{\binom{n_k}{m_k}B(a_{k-1},b_{k-1})r^{a_{k-1}+m_k-1}(1-r)^{b_{k-1}+n_k-m_k-1}}{\binom{n_k}{m_k}B(a_{k-1},b_{k-1})\int_0^1 r^{a_{k-1}+m_k-1}(1-r)^{b_{k-1}+n_k-m_k-1}dr} =$$

$$= \frac{r^{(a_{k-1}+m_k)-1}(1-r)^{(b_{k-1}+n_k-m_k)-1}}{B(a_{k-1}+m_k, b_{k-1}+n_k-m_k)}$$

$$\sim Beta(a_{k-1}+m_k, b_{k-1}+n_k-m_k)$$

$\square$

www.ingramcontent.com/pod-product-compliance
Lightning Source LLC
Chambersburg PA
CBHW060444240326

41598CB00087B/3482